Practical Pneumatics

Practical Pneumatics

Chris Stacey

Routledge
Taylor & Francis Group

LONDON AND NEW YORK

First published by Newnes

This edition published 2011 by Routledge
2 Park Square, Milton Park, Abingdon, Oxon OX14 4RN
711 Third Avenue, New York, NY 10017, USA

Routledge is an imprint of the Taylor & Francis Group, an informa business

First published 1998

British Library Cataloguing in Publication Data
A catalogue record for this book is available from the British Library

Library of Congress Cataloguing in Publication Data
A catalogue record for this book is available from the Library of Congress

ISBN 978-0-415-50295-5

Contents

Note

Further material available from the author:

Model solutions for the exercises which follow each chapter.

OHPs or files for presentation in PowerPoint ® format for all the line drawings featured in this book.

Apply to the author, care of the publisher.

Preface

The City and Guilds 2340 Scheme in Fluid Power Engineering Part 2 (Pneumatics) is one of a series of programmes of study for pneumatics and hydraulics developed by the Education and Training Committee of the British Fluid Power Association using the City & Guilds of London Institute as the awarding body. Practical competence is the main objective and success results in the award of the Certificate in Fluid Power Engineering for the appropriate subject and grade. Assessments to determine whether a candidate has the required competence and knowledge use, where possible, the methods developed in National Vocational Qualifications in which success is judged according to evidence of competence rather than by traditional teaching and testing, but since competence in a technical occupation must be based on underpinning knowledge, some theoretical instruction and assessment is essential. The course offered by an approved training centre will comprise all the elements of competence and knowledge in a flexible format so that a candidate with evidence of prior experience in a relevant field may arrange to omit parts of the training.

This book results from the author's experience researching course material and setting up practical tasks for the Technology Training Works of Frizington, Cumbria. Grateful acknowledgement is made to Mr Ian Stephenson, who created and equipped this training facility, for his support and assistance and to the students who helped to develop the course material on which much of this book is based. The aim has been to present all the requirements under one cover and to convey the principles so that they can be readily understood by the average candidate who is more at home with practical concepts than with theory. The subjects are approached progressively to be learned in the order presented, and as some are spread over a number of chapters, to assist the student the syllabus is reproduced in Appendix A with each element referred to page numbers. Most chapters conclude with questions to help the development of underpinning knowledge. Some components and systems which fall outside the scope of the syllabus or which are fully covered in the Part 3 Syllabus 'Electrical/Electronic Control of Fluid Power Systems' are covered in outline only.

Pneumatics is the means by which man has used the most abundant fluid in nature to do work for many centuries, and is now less expensive, more sophisticated and easier to purchase and apply than ever before. Used alone or together with electrical equipment or electronics, it provides a formidable tool to create wealth. It is hoped that this book will help to show those who use or maintain pneumatic equipment how interesting and rewarding fluid power engineering can be, and in conjunction with the City & Guilds competence based programmes promote greater understanding and appreciation of the valuable potential of pneumatics for the 21st century.

Finally, my thanks to Alf Leatherbarrow and John Savage, of the British Fluid Power Association's Education and Training Committee, for bringing their considerable experience to bear on reading the text and technical presentations and for their corrections and constructive criticism.

Chris Stacey
Cumbria, October 1997

1
Practical skills

This book, while applicable to several types of pneumatics course, is targeted in particular at the City & Guilds 2340 scheme in Fluid Power Engineering Part 2 (Pneumatics). Basic practical skills are generally acquired in the working environment, but this programme of study recognizes the need for further training in the understanding and application of pneumatic technology, together with regard to safety, project planning and systematic maintenance and fault-finding.

Underpinning knowledge is included here only in so far as it is necessary to support the practical skills; however, to support the level of skill which is expected and necessary in today's workplace, a considerable theoretical syllabus, reproduced in Appendix A, has resulted.

1.1 Aims of the City & Guilds 2340 Scheme in Fluid Power Engineering

The basis, aims and products of the City & Guilds programme are comprehensively covered by the *Scheme pamphlet*, together with *Tutors/Trainers – Notes for Guidance*, obtainable from City & Guilds, London. To quote briefly and selectively from the scheme pamphlet:

Aims of certification
Candidates who achieve the certificate(s) in Fluid Power Engineering Competences will have demonstrated that they have acquired

(a) an ability to perform competently practical tasks relevant to the installation, commissioning, maintenance and fault-diagnosis of pneumatic systems.

(b) a generalised practical mastery of the technology used in their practical tasks so that they may progress to other applications in new tasks or new training without relearning the main skill content of the process.

(c) the necessary competences in practical communication, task planning, doing and checking the results of work.

(d) a basis for informed assessment of their personal aptitudes and attitudes in relation to their work.

(e) confidence in a new role.

Practical tasks and assessment – aims

The aim of assessment is to provide evidence that candidates have acquired practical competence. The criterion of success is the demonstration of the ability to do the job and mastery criteria are used to determine how well candidates perform. These criteria must be met for candidates to be considered competent.

The programme is structured so that the existing skills of a candidate, based on evidence provided, are recognized as elements contributing to a final certificate. Quoting again from the scheme pamphlet:

> It is stressed that certificates are awarded for the acquisition of competences whether in an integrated course or by credit accumulation and not for serving a period of study. It is for this reason that no reference is made in the scheme to fixed periods of study time or how long a course should be: this will depend on the background and abilities of trainees undertaking the course.

1.2 Ability criteria

The ability criteria have been devised as a framework for the teaching and assessment of skill and split into the following four categories:

1.1 Interpret pneumatic circuit diagrams
1.2 Construct pneumatic systems from given information
1.3 Identify and rectify faults in pneumatic systems
1.4 Carry out routine maintenance on pneumatic systems.

Each category is broken down further into the elements of competence:

1.1 1.1.1 Components correctly identified.
1.1.2 Application of components identified.
1.1.3 Operation of pneumatic system relating to control inputs and machine outputs identified.
1.2 1.2.1 Appropriate components selected and adjusted as necessary.
1.2.2 System assembled in a safe and efficient manner.
1.2.3 Start-up and commissioning procedures correctly specified and followed.
1.2.4 System operates according to requirements.
1.2.5 Safe working practice and statutory regulations followed at all times.
1.3 1.3.1 Nature of faults correctly identified.
1.3.2 Fault-finding checklist prepared.

1.3.3 Diagnostics used to locate fault, ensuring safety at all stages.

1.3.4 Machine/system shut down safely in correct sequence as necessary.

1.3.5 Faulty component repaired/ replaced/adjusted as necessary.

1.3.6 Cause and effect of faults correctly assessed.

1.3.7 Machine/system recommissioned in accordance with set procedures.

1.3.8 Machine/system operates according to requirements.

1.3.9 Safe working practices and statutory regulations followed at all times.

1.4 1.4.1 Service/maintenance require established schedule.

1.4.2 Servicing/maintenance undertaken, as per schedule, in safe and efficient manner.

1.4.3 Performance testing of components carried out, as necessary.

1.4.4 System tested after maintenance to ensure efficient working.

1.4.5 Safe working practice and statutory regulations followed at all times.

1.3 Elements of competence

Candidates may be understandably put off by the dry descriptions of skills required, so here is a summary which may help:

1.1 Interpret pneumatic circuit diagrams
— You must recognize components and know how to apply them.
— You must understand how a circuit is controlled and how it carries out its working functions.

1.2 Construct pneumatic systems from given information
— You must be able to select and correctly adjust components to perform the working functions of a system, assemble them safely and efficiently, and commission the system using safe procedures. The system must work as specified and you must follow safe practices and be aware of and observe the applicable safety regulations.

1.3 Identify and rectify faults in pneumatic systems
— You must identify the faults in a system correctly, prepare a checklist to provide a systematic fault-finding procedure and locate the faults with diagnostic testing with regard to safety throughout. A safe shutdown sequence must be used. The circuit or component faults must be repaired or corrected by adjustment or replacement and the cause and effect of faults identified. The machine must be started up using a safe procedure, operate as specified, and you must follow safe practices and be aware of and observe the applicable safety regulations.

1.4 Carry out routine maintenance on pneumatic systems
— An established schedule must be used for all servicing/maintenance, which must be carried out as scheduled, safely and efficiently. Components and systems should be performance tested after maintenance. You must follow safe practices and be aware of and observe the applicable safety regulations.

In the training environment, practical assessment tasks may be small-scale systems created on a 'pegboard' or in a training unit providing a selection of small components. Safety is a recurrent feature of the assessment criteria to reflect the considerable danger associated with pneumatics in factory applications. Training rigs seldom duplicate these dangers, so normally they must be imagined, and all procedures carried out and documented as if the system is large, powerful and dangerous.

2
Units and calculations

2.1 Units of measurement

The SI metric **base units** have a consistent logical basis: all measurements are made in the terms of a basic unit applicable to the property. If this results in awkward figures, then prefixes can be applied which multiply or divide the basic unit by 10 or factors of 10 as required. Grams, for example, which are a small unit of mass or weight, are more often encountered in ordinary life as kilograms, i.e. units of 1000 grams, but for tiny measurements they are also encountered as milligrams, i.e. units of $\frac{1}{1000}$ of a gram.

The prefixes are similar for all the basic units. The more common ones are given in Table 2.1.

Table 2.1 Common unit prefixes

Prefix name	Symbol	Factor
mega	M	× 1 000 000
kilo	k	× 1000
deca	da	× 10
deci	d	× 0.1
centi	c	× 0.01
milli	m	× 0.001
micro	μ	× 0.000 001

The basic units used in pneumatics are as shown in Table 2.2 – imperial equivalents (in brackets) are for comparison only.

Other practical pneumatics units are **derived units** or combinations of basic units, some with special names due to popular usage, to which the prefixes do not apply. Force, for example, is derived from mass, a measurement of the effect of a mass subject to gravity. Pressure is a combination of force and area (force/unit area).

These other units used in pneumatics are given in Table 2.3.

Table 2.2 Basic units used in pneumatics

Property	Unit name	Symbol	Equivalent to:
Mass or weight	Kilogram	kg	(2.2 lb)
Length	Metre	m	(39.37 in)
	Millimetre	mm	0.001 m
	Decimetre	dm	0.1 m
	Micrometre ('micron')	μm	0.001 mm (unit of filtration)
Area	Square metre	m²	
	Square millimetre	mm²	
Time	Second	s	
	[longer times use minute (min) and hour (h)]		
Power	Kilowatt	kW	(1.34 horsepower)

Table 2.3 Derived units used in pneumatics

Property	Unit name	Symbol	Equivalent to:
Force	Newton	N	(0.22 pound force)
Pressure	Bar	bar	0.1 N/mm² (14.5 lb/in²)
Volume	Litre	l	1 dm³ (0.035 cu.ft)
	Cubic metre	m³	1000 litres (35.3 cu.ft)
Flow	Litre/minute	l/min	0.06 m³/h (0.035 ft³/min)
	Cubic metres/hour	m³/h	(0.588 ft³/min)
Velocity	Metre/sec	m/s	

2.2 Standard form notation

Very large or very small numbers written in conventional notation are not easy to read. As examples:

- The force resulting from gravity acting on a mass of 5 tonne is 50 000 newton, or 5.0×10^4 newton in standard form notation (note that large numbers use $10^{positive\ power}$).
- A fine filter particle diameter standard of 0.3 micrometre (0.3 μm) could be written as 0.0003 mm or 3.0×10^{-4} mm in standard form notation (small numbers use $10^{negative\ power}$).
- Pascal (Pa) is the SI unit of pressure equal to 1 N/m², little used outside meteorology because it is so small, e.g.

$$8.5\ bar = 850\ 000\ Pa \qquad or \quad 8.5 \times 10^5\ Pa$$
$$8.5\ pascal = 0.000\ 085\ bar \qquad or \quad 8.5 \times 10^{-5}\ bar$$

The simple rule for translating numbers to and from standard form notation is to move the decimal point – the number of places moved is the power of 10:

- from standard form to conventional notation:
  — if the power of 10 is positive move the point to the right
  — if the power of 10 is negative move the point to the left
  — then add noughts in each vacant place;
- from conventional to standard form notation:
  — for large numbers move the point left until one digit remains to the left of the point
  — for small numbers move the point right until there is one digit with a value other than zero to the left of the point
  — then remove surplus noughts.

In most pneumatics applications, it will be possible to avoid very large or very small numbers by the use of units with suitable prefixes.

2.3 Areas and volumes

Here are some brief reminders of how areas and volumes are calculated (Figs 2.1–2.5).

The area of a circle is calculated using the geometrical value π (= 3.142).

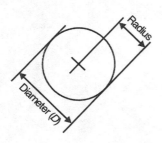

Fig. 2.1 Circle

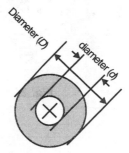

Fig. 2.2 Annulus

The area of a circle $= \dfrac{\pi \times D^2}{4}$

The area of an annulus $= \dfrac{\pi \times (D^2 - d^2)}{4}$

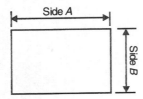

Fig. 2.3 Rectangle

The area of a rectangle or square $= A \times B$

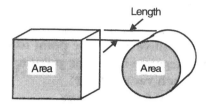

Fig. 2.4 Cube or cylinder

The volume of a cube or cylinder = area × length

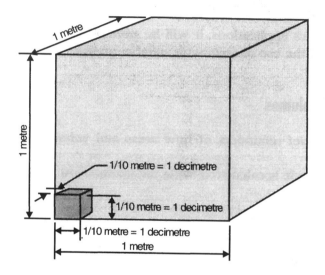

Fig. 2.5 1 litre = 1 cubic decimetre (dm³)

3
Properties and behaviour of air (1)

3.1 Pressure

All fluids can convey force from one place to another. Remember that fluid can be liquid or a gas such as air. A force applied to the surface of fluid in a container will cause a force to occur at every point where the fluid is contained. These forces, which result from the fluid trying to escape, apply in all directions.

Pressure is the measure of how much force is present on an area of the fluid. The pressure is created by applying the force to an area, so a force applied to a small area will give a high pressure, while the same force applied to a large area will give a low pressure. When the pressure in the fluid has conveyed the force through the fluid to another location, the force is available for use. A high pressure in the fluid acting on an area will create a large force, or a low pressure acting on the same area will create a small force. Also, a low pressure applied over a large area results in a large force, and a high pressure applied over a small area will create a small force.

> **Definition of pressure for pneumatic purposes**
>
> Pressure is the force in air due to compression, per unit of surface area

To calculate pressure created in a fluid by the application of a force:

$$\text{pressure} = \frac{\text{force}}{\text{area}}$$

To calculate force exerted by pressure acting on an area:

$$\text{force} = \text{pressure} \times \text{area}$$

The units (see Ch. 2) are as follows:

- Force – newton (N)

- Area – mm²
- Pressure – bar.

To apply the relationships above, use bar/10 for pressure, i.e.

$$\frac{\text{pressure (bar)}}{10} = \frac{\text{force (N)}}{\text{area (mm}^2)}$$

and

$$\text{force (N)} = \frac{\text{pressure (bar)}}{10} \times \text{area (mm}^2)$$

Liquids and gases behave differently because in non-control situations liquids are considered to be incompressible where gases are compressible. When force is applied to a liquid, the density of the liquid does not change: the force must remain applied for the liquid to retain pressure, so when the force is removed the pressure disappears.

Force applied to a gas (such as air) changes the density of the gas: if the compressed gas is contained, as shown for example in Fig. 3.1, then the pressure in the gas remains and is available for use after the original force has been removed.

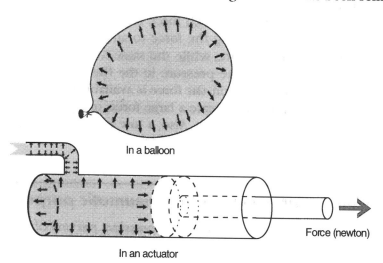

In a balloon

In an actuator

Force (newton)

Fig. 3.1 An explanation of pressure

Worked example 3.1

Calculate the force (neglecting frictional losses) resulting from a pressure of 6 bar acting on a piston area of 100 mm².

Force (N) = pressure (bar)/10 × area (mm²)
so force = 6/10 × 100 = 60 N   Ans

3.2 Alternative ways of measuring pressure

A deep-water fish lives in water under pressure: we live in compressed air. The air around us has been compressed by the Earth's gravity acting on the atmospheric layer, so the air we compress for use in industry is therefore already part-compressed. Weather forecasters describe the varying atmospheric pressure in millibar. (Chapter 2 explains how the ISO system of units uses prefixes: millibar means bar/1000 or bar $\times 10^{-3}$.) For industrial use we can normally overlook the exact and varying atmospheric pressure of air and assume that it is 1000 millibar or 1 bar, and furthermore we can discount it totally and make all our working measurements of air pressure on a base of atmospheric pressure, so that as far as our pressure gauges are concerned, atmospheric pressure = 0 bar. Measurement of pressure using the **gauge pressure** standard, therefore, gives values 1 bar less than the same pressures measured against the **atmospheric pressure** standard.

Gauge pressure provides us with figures which we can apply in all the calculations required in most industries, except for one common calculation: the change in volume of air due to compression (see Section 3.3).

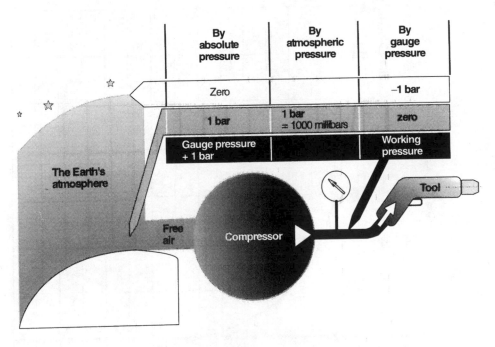

Fig. 3.2 The three alternative ways of measuring pressure

3.3 Changes when air is compressed

Air volume varies inversely with pressure: when the pressure doubles the volume halves or when the pressure halves the volume doubles, and so on. This simple relationship is only true, as one might expect for such a natural extension of the behaviour of atmospheric air, using the atmospheric scale of measurement and not the industrial gauge scale of measurement. For use in volume calculations, the atmospheric scale is renamed the **absolute** scale of measurement since the process started in outer space. There are therefore three common alternative standards of measurement, as shown in Fig. 3.2, the atmospheric scale being mainly for weather buffs and not for industrial plant. Figure 3.3 shows graphically how air volume changes with the pressure applied and how gauge pressure relates to absolute pressure.

Other changes happen to air when it is compressed, which we have no doubt noticed when using a bicycle pump: it becomes hotter and (as observed from the interior of a cycle tyre inner tube) it sheds moisture. These changes are important for air used in industry because they are a major influence on the quality and cost of compressed air. The heat in air causes considerable but temporary expansion of the air, resulting in increased handling costs and a loss of volume as it cools during distribution, and so must therefore be minimized. Depositing water from air into industrial machinery could cause damage, so we must be able to calculate the capacity of dryers that would remove it.

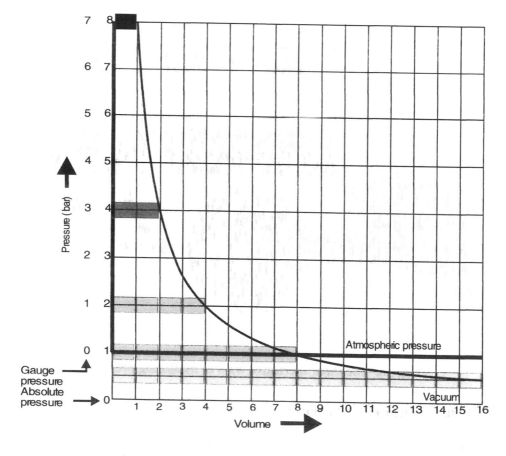

Fig. 3.3 How air volume changes with pressure

It would surely be difficult to find a medium for the transmission of energy which is cheaper than air! However, compressed air is in practice very expensive due to the low efficiency of compression and distribution, so small improvements in the design of compressors or air treatment plant soon pay for themselves. Since heat is the main product of losses during compression, the design of compressors to generate cool air is a closely studied and continually evolving science. Remembering how the bicycle pump heated the air, consider the difference between (a) a pump with a thin steel cylinder and (b) a pump with a double-skinned plastic cylinder. In both cases the process of compression would generate the same heat, but whereas with (a) most of the heat would be dissipated, in (b) most of the heat would remain trapped in the air. Pump (a) would generate relatively cool air, but most of the heated air in (b) would enter the tyre where it would cool and therefore lower the tyre pressure, while the remainder of the heated air would remain in the pump and add to the heat generated during the next stroke. This explains the main reasons why compressors with the best ability to dissipate heat are more efficient.

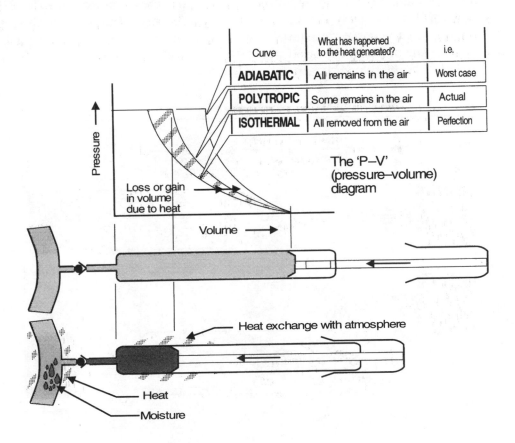

Fig. 3.4 The three standards of compression

Figure 3.4 shows, in the same graphical way as in Fig. 3.3, how air volume would change as it is compressed by the bicycle pump illustrated under the **PV diagram.** It also shows the effect of heat on the air volume, together with the Greek names which are given to the worst, the best and the typical in-between standards of compressor performance.

3.4 Flow

Flow occurs between two points only if one point starts at a higher pressure than the other point. Consider two rooms separated by an airtight door. In one of the rooms an air bottle is vented to raise the air pressure in the room. When the door is opened, air will flow from the room with high pressure into the room with low pressure. Pressure therefore causes flow. If the pressures were the same, there would be no flow.

However, flow cannot cause pressure. Pneumatics course candidates often confuse these two properties of an air system, which for most applications can be regarded as independent. An air jet may appear to have pressure because of its high rate of flow; in reality the jet should be regarded as an extension of the pipe continuing to convey the pressure imparted into the air during compression, but with a high rate of leakage to atmosphere. The jet must strike the workpiece before the leakage has dissipated the pressure.

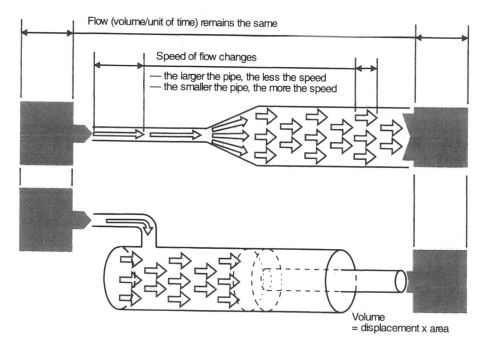

Fig. 3.5 An explanation of flow

Consider a liquid flowing through a system of varying pipe diameters. A liquid is chosen instead of a gas because the density remains the same and so the behaviour is simpler. Three rules apply:

1. All liquid that goes in at one end must come out at the other end, so the quantity of flow is constant through all pipes whatever the diameter.
2. The speed of flow, which is not to be confused with the quantity, changes in inverse proportion to the pipe area, as illustrated in Fig. 3.5.
3. Discounting the relatively small pressure difference which has caused the flow to occur, raising the pressure throughout the system makes no difference to the flow.

Air would behave in just the same way as the incompressible liquid provided the pressure and temperature remain the same, but under all practical circumstances both pressure and temperature vary sufficiently to alter the volume of the compressible air, or to raise its density which is the same thing. Therefore for air, the quantity of flow at any point varies in inverse proportion to the absolute pressure applicable at that point.

Definition of flow for pneumatic purposes

Flow is the volume of air passing a point, per unit of time

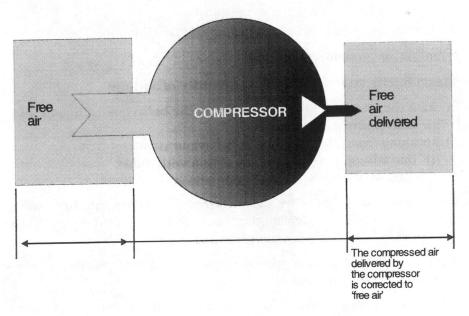

Fig. 3.6 What is 'free air delivered'?

3.5 Free air

Since air under pressure varies in volume, to compare quantities of air, for example when making a comparison between two alternative makes of compressor, a standard measurement is required, so an air volume is converted to the equivalent **free air** volume, that is to what its volume would be at 1.013 bar$_{abs}$ and 15°C. The correction is a simple application of two factors: one as described in Section 3.3 to correct for the change in absolute pressure, and the other an adjustment to correct for the expansion of the air according to temperature. Compressor capacities are quoted by manufacturers as their **free air delivered** capability or **FAD**.

Definition of free air

A volume of air considered to be at mean atmospheric pressure of 1.013 bar$_{abs}$ and temperature of 15°C

Definition of free air delivered

The volumetric flow rate of a machine or system, expressed in terms of standard free air

---

Exercises for Chapter 3

1. State what is meant by free air delivered (FAD).   2.3(a)

2. Define (a) air pressure, (b) air flow.   2.3(a)

3. Explain briefly **velocity** of air flow, **quantity** of air flow.   2.3(h)

4. (a) State the essential difference between the behaviour of liquids and gases
      for fluid power applications.
   (b) Resulting from the difference in behaviour, name:
      (i) one advantage of gas in comparison with liquid;
      (ii) one disadvantage of gas in comparison with liquid.   2.3(h)

5. For a pneumatic cylinder with an air pressure of 6 bar applied to the full piston area of 100 mm², calculate the theoretical force on the piston. (Show all your working. Make no allowances for losses.)   2.5(a)

6. State the principal reason why the absolute scale of pressure is sometimes employed for air calculations.   2.3(b)

7. If the gauge pressure reading for a system is 7.0 bar, state the corresponding absolute pressure for the air in the system.   2.3(b)

8. If the atmospheric pressure is exactly 1000 millibar (= 1 bar), state the reading shown on unloaded gauges (gauges with no system pressure applied) which are graduated to read: (a) gauge pressure, (b) absolute pressure.   2.3(b)

9. Name, describe briefly and state the difference between:
   (a) the scales used for the pressure of air in industrial equipment, and
   (b) the ambient pressure quoted in weather reports.   2.3(b)

10. Of two alternative types of air compressor design – (a) a compressor with no heat exchanger and well insulated casing, and (b) a compressor with no insulation and an effective heat exchanger – which will give the more efficient installation?   2.4(c)

11. Sketch a typical $P-V$ (pressure–volume) diagram for the compression of air showing which axis of the diagram represents change in volume and which represents pressure, with three approximate compression curves shown: the adiabatic (worst case); the polytropic (typical actual situation): and the isothermal (perfect performance).   2.3(c)

12. A totally leak-free air distribution system receives free air delivered of 1000 l/min from a compressor. What is the total quantity of air delivered by the system to the plant when corrected to free air?   2.3(h)

13. As air travels through pipework, changes in pipe diameters result in changes to the velocity of air flow. If, for a steady quantity of air delivered, the flow velocity is 5 m/s through a pipe with an area of 1000 mm², what will the air velocity be if the pipe area reduces to 500 mm²? (Assume that the pressure, and therefore the density of the air, remains constant.)   2.3(h)

4
Factory air service systems

4.1 System requirements

Compressed air for factory use must first be conditioned to ensure that it is clean, dry and cool before it is distributed to the working locations. Air plant, like other fluid power systems, can deliver useful results even when badly designed, and poor quality, inefficient compressed air systems are not uncommon. However, careful attention to every feature of a system to give maximum efficiency and minimum maintenance costs little more but results in much reduced running and maintenance costs.

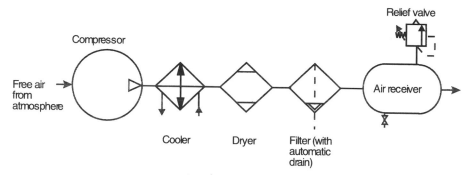

Fig. 4.1 Air treatment

Every component can be represented by a symbol to the ISO1219 standard. Figure 4.1 shows diagrammatically the typical process of air compression, treatment and storage. Sometimes the storage vessel or 'air receiver' is located before the treatment components.

4.2 Compressors

Most factory compressors are of the **displacement** type, in which force is applied to a contained volume of air which is then discharged. **Dynamic** type compressors based on the fan or turbine principle are sometimes fitted for high flows and low pressures.

Compressor capacity is described in terms of the flow and pressure attainable. Flow is usually stated in metre³/hour (m³/h) of free air (free air is explained in Ch. 3): 1 m³/h = 16.67 l/min. Pressure attainable is given in bar. The free air output is often tabulated or charted over a range of working pressures, since the efficiency is lower at higher pressures and therefore the output is less. Figure 4.2 shows the typical capabilities of various types of compressor.

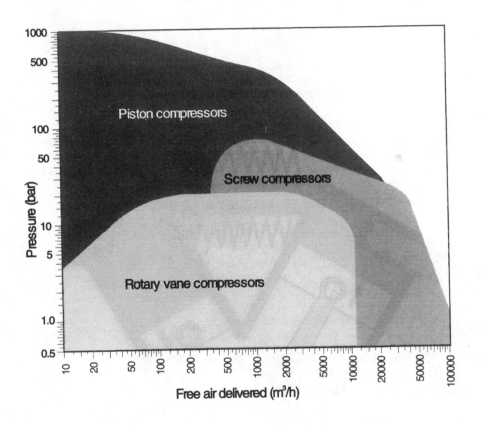

Fig. 4.2 Compressor types – typical scope

Piston compressors can deliver compressed air to suit the requirements of any system. They are typically noisier than screw or vane compressors and the noise can be conveyed by the air in the form of pulsations in the supply, single cylinder types being by far the worst in this respect (Fig. 4.3). Multistage piston compressors are more efficient, mainly because of 'intercooling' between the stages, and can be capable of higher maximum pressures (Fig. 4.4).

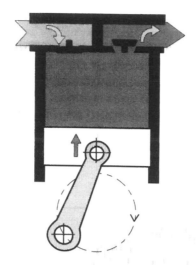

Fig. 4.3 Single stage piston compressor

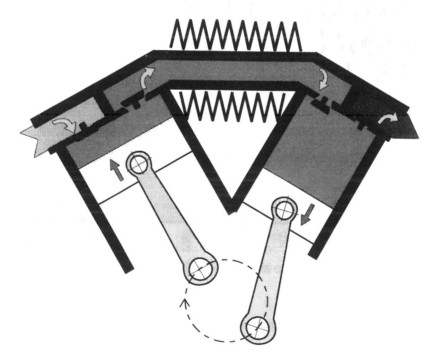

Fig. 4.4 Multistage piston compressor

A feature common to all multistage compressors is the reduction of piston arca in inverse proportion to the pressure. This balances the crankshaft forces, minimizes the torque required to drive the compressor and so contributes to efficiency. Intercoolers may be air- or water-cooled.

Rotary vane compressors are quiet and inexpensive (Fig. 4.5). To maintain lubrication of the vane edges, recirculating lubrication systems are sometimes required. Multistage versions include coolers between the stages.

Fig. 4.5 The vane compressor

Screw compressors are quiet and efficient but expensive, so they are more frequently chosen for larger systems (Fig. 4.6). A rotor with male lobes meshes with a smaller diameter rotor with female lobes. Oil lubricates the bearings, seals the clearances and cools the air before being purged from the air at discharge.

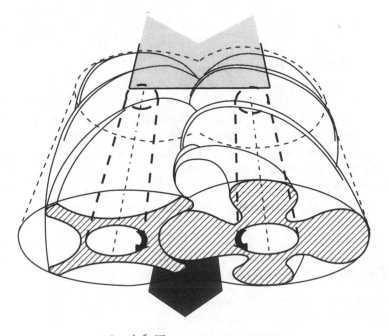

Fig. 4.6 The screw compressor

Diaphragm compressors are an inexpensive reciprocating type able to deliver oil-free air for very small systems (Fig. 4.7). Factory applications would typically be as a stand-alone supply for small-scale functions within a machine.

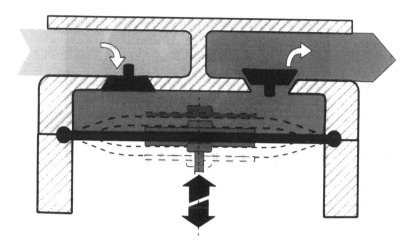

Fig. 4.7 The diaphragm compressor

4.3 Volumetric efficiency

To state the volumetric efficiency of a compressor, the volume discharged is corrected to free air (and is therefore equal to the free air delivered by the compressor) and the corrected volume is then compared with the free air ingested by the compressor. Because of losses, less air is discharged than is ingested. The ratio is expressed as a percentage (see Fig. 4.8).

The volume of air ingested can be considered as equal to the swept volume (area of bore × stroke) or **displacement** of the compressor.

Definition of volumetric efficiency

The free air delivered divided by the compressor displacement, expressed as a percentage

---

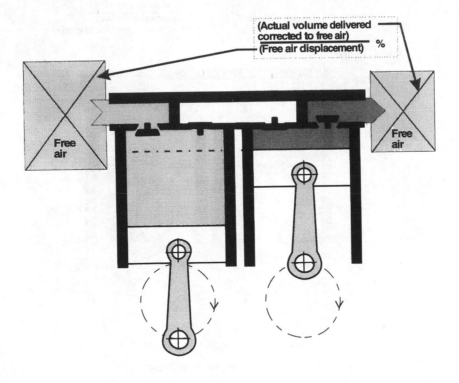

$$\frac{\text{(Actual volume delivered corrected to free air)}}{\text{(Free air displacement)}} \%$$

Free air

Free air

Fig. 4.8 Compressor efficiency

4.4 Compression ratio

The compression ratio of a compressor is a comparison between the free air ingested (or displacement) and the actual volume of compressed air discharged at the maximum attainable pressure (Fig. 4.9). This is most simply represented (since volume and pressure vary in inverse proportion) by the maximum discharge pressure measured according to the absolute scale.

Definition of compression ratio

The maximum pressure the compressor can deliver related to atmospheric pressure (i.e. as measured in bar$_{abs}$)

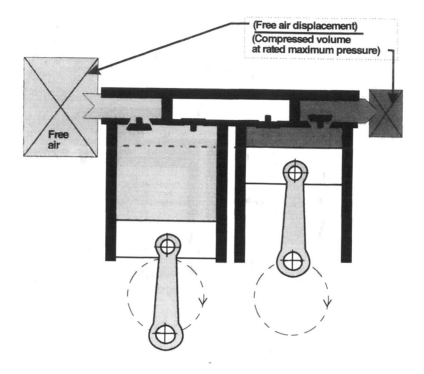

(Free air displacement)
(Compressed volume at rated maximum pressure)

Free air

Fig. 4.9 Compression ratio

4.5 Air treatments

Cooling of the air is carried out between stages of compression and after compression by means of heat exchangers which vary in type from simple air-to-air finned arrangements, through recirculating water exchangers to sophisticated cooling towers where cooling is achieved through continuous vaporization of water. Cooling is also frequently carried out during the drying process.

Drying of compressed air is probably the most important form of air treatment because water contamination results in inefficiency and can cause serious damage to equipment. **Refrigeration dryers** incorporate a refrigeration unit like a domestic refrigerator to cool the air to a temperature close to freezing which precipitates the entrained water. Heat is then exchanged with the warm incoming air. This both cools the incoming air and, at the same time, warms the air discharged to normal ambient temperature (Fig. 4.10).

The **absorption dryer** is a container through which the air is passed incorporating a layer of replaceable drying medium (Fig. 4.11). Vanes around the air inlet encourage the air to rotate. This is a common feature of air treatment devices which throws out some of the moisture and contaminants. Water can also drop from the drying chemical into a drained well beneath.

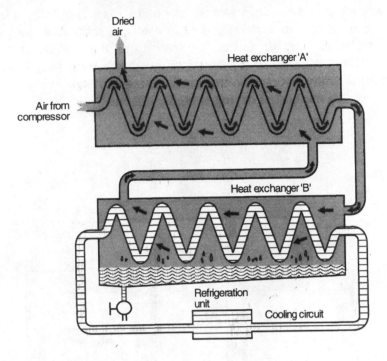

Fig. 4.10 The principle of the refrigerant dryer

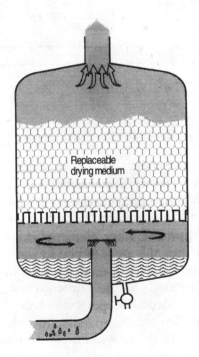

Fig. 4.11 The principle of the absorption dryer

The **adsorption dryer** comprises twin containers each holding a layer of non-replaceable honeycomb material (Fig. 4.12). Valves divert the main air path through one container to dry the air and at the same time divert some of the dried air through the other container to dry the medium. After a short period, all valve positions are automatically reversed so that each container alternately dries the air and is regenerated. This method of drying is highly effective.

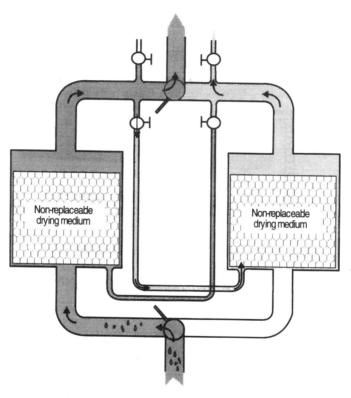

Fig. 4.12 The principle of the adsorption dryer

Filters are fitted before and after distribution of the air (Fig. 4.13). The standard grade employed for typical factory applications has a typical nominal filtration standard of 40 μm for main line application or 5 μm for local applications, that is it removes particles down to a diameter of 40 or 5 micron (a micron is 0.001 mm). For specially sensitive equipment, filter elements down to 0.1–1 μm rating are available. On entering the filter, the air is forced to rotate so that heavier particles are thrown out. To exit the filter the air must pass through the element which is a porous screen. Element materials include sintered metals and plastics, sometimes cleanable by washing, or bonded fibre types which are normally replaced for maintenance. A baffle beneath the element separates the circulating air from water and contaminants in the bottom of the bowl. A drain enables these to be removed during routine maintenance. To enable the plant to continue working, main line filters are often duplicated and switched into circuit by changeover valves.

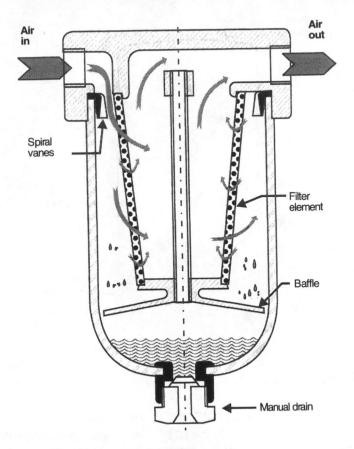

Fig. 4.13 The principle of the filter

Lubrication is sometimes part of the main line treatment installation if the entire plant relies on lubricated air. Lubricators are described under 'service units'.

Oil removal is more frequently part of the main line treatment installation since few compressors generate totally oil-free air, as modern equipment increasingly requires no lubrication and because of concerns about the health risks from oily air released into the atmosphere. However, removal of oil by the use of special filters at every exhaust outlet throughout a plant is more common.

The **air receiver** is a pressure container fitted in all systems to store air, in order to cater for variations in demand and to smooth out compressor pulsations from the supply (Fig. 4.14). The compressor is controlled by a pressure switch, with the 'on' setting about 0.75 bar below the 'off' setting. In case for some reason this system of pressure regulation should fail, the receiver is always fitted with a relief valve. Some moisture and sludge will continue to settle out of the air in the receiver, so an automatic drain trap is always fitted. A visual or remotely monitored thermometer is also usual. An inspection hatch is essential for cleaning and inspections under the pressure systems regulations.

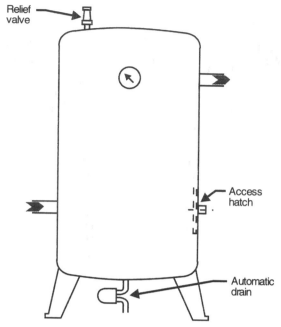

Fig. 4.14 Typical air receiver

4.6 The automatic drain valve

Automatic drains are found on most filters or air receivers and are of many different designs. To enable scum and sludge to be discharged, the drain orifice must not be too small, but to ensure the device responds to small quantities of contaminants the float must not be too large. Most designs use air pressure, switched by a small float-operated poppet, to actuate the discharge valve (Fig. 4.15).

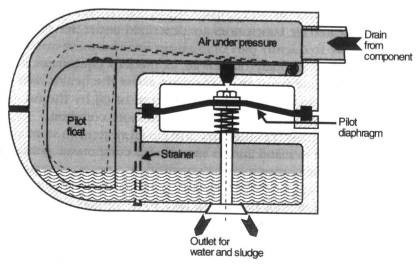

Fig. 4.15 The principle of the automatic drain

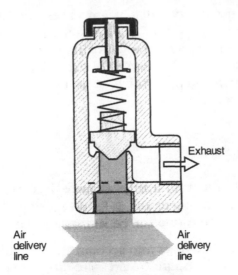

Fig. 4.16 The direct acting relief valve

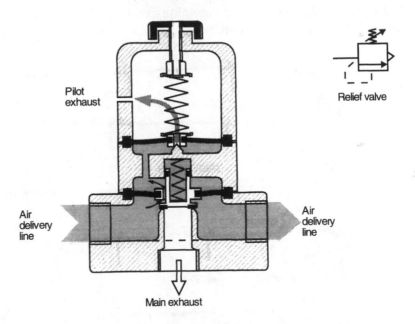

Fig. 4.17 The pilot operated relief valve

Relief valve

4.7 The relief valve

A relief valve branches from a pressure line to spill off pressure which exceeds the setting of the valve to atmosphere. Simple **direct acting** types comprise a poppet and spring (Fig. 4.16). The venting pressure rises as flow increases and they can be noisy in operation.

Pilot operated types use two diaphragm valves to give constant venting pressure with quiet operation (Fig. 4.17). A small valve opens if line pressure exceeds the spring setting. A tiny orifice through the large valve diaphragm allows a little air into the space between the diaphragms, resulting in a pressure imbalance across the large valve diaphragm which opens and remains open until line pressure equals the relief valve setting

4.8 The distribution network

After compression, treatment and storage, air must be carefully distributed by a system designed to cope with present and, as far as possible, future requirements (see Fig. 4.18). The main considerations can be summarized as follows:

- Ensure that pipe diameter takes into account the length and complexity of the system and results in an acceptable pressure drop when the flow is at the maximum value anticipated.
- Design a network that will accept new connections and extensions.
- Use a ring main and cross-connections to ensure adequate flow to all users, allowing for present and anticipated peak demands.

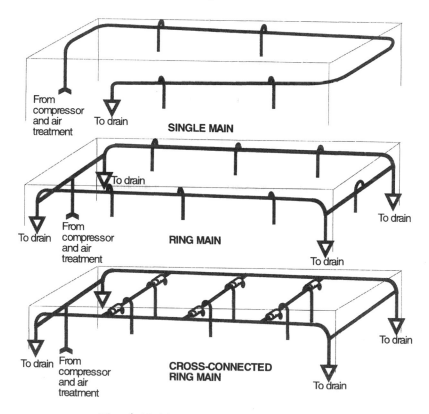

Fig. 4.18 Types of compressed air main

- Fit isolation valves for each separate workshop system, level or major tool location.
- Fit isolation valves to permit the removal of any major component without shutting down the system.
- Ensure all pipework is accessible for maintenance and cleaning.
- Allow for expansion with clearances on all mountings and supports.
- Attach draw-off points from the top of the loop, and ensure all pipework falls towards drain valves and that every low point can be drained.

4.9 Air treatments at point of use

After distribution, additional air treatment and other components are normally positioned at the point of use. These are:

- An isolation valve – an on–off valve for the plant served at this point.
- A filter – similar to the filter already described with very probably a higher (smaller particle size) filtration rating.
- A pressure regulator and pressure gauge. The regulator is an adjustable component which allows mains pressure downstream of the plant only up to the pressure to which it has been adjusted. It therefore imposes a limit on the plant pressure. A secondary function of most regulators is to vent pressure from downstream (i.e. from the plant) when that pressure exceeds the regulator setting – a situation which can arise (a) if the regulator setting is reduced or (b) if actuator loads cause back-pressures.
- A soft-start/dump valve (if required).
- A lubricator (if required).

All these components are typically grouped together in an 'FRL' – Filter-Regulator-Lubricator assembly, or 'air service unit' (Fig. 4.19). The design

Service unit
(FRL)
Symbol to ISO1219

Fig. 4.19 A typical FRL assembly. (Photograph by kind permission of CompAir Maxam Ltd.)

enables subdivision of the assembly, so that, for example, if a lubricator is not required, it need not be included, and also so that removal of the component for maintenance or replacement can be carried out without breaking down pipework. A drain valve, sometimes incorporated in a 'drip leg drain' component, is also fitted adjacent to the FRL at the lowest point of the supply leg.

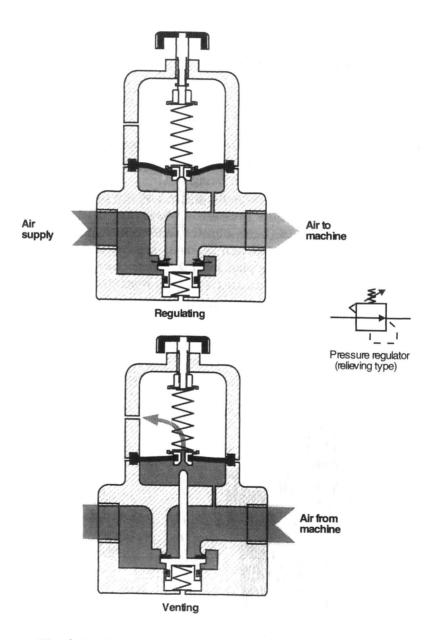

Fig. 4.20 The principle of the pressure regulator

The regulator reacts only to pressure from downstream (from the machine) (Fig. 4.20). When **regulating**, the diaphragm centre orifice remains closed. Spring pressure (according to adjustment) is balanced by the pressure acting under the diaphragm – if it is less the poppet opens to allow flow to the machine; if it is more the poppet closes.

When venting, the poppet first closes, because downstream pressure has exceeded the setting. The diaphragm continues to lift, which opens the diaphragm centre orifice to allow air to vent into the bonnet.

The **soft-start/dump valve** (Fig, 4.21) is a safety device to give gradual pressurization of a system at start-up. If, due to an emergency or uncontrolled shutdown, the system is not at rest, then the soft-start valve will ensure that actuator movements are not violent. The valve timing is adjustable and a solenoid control can be linked to sensors on the machine.

Fig. 4.21 A typical air service unit with soft-start valve. (Photograph by kind permission of SMC Pneumatics (UK) Ltd.)

The **lubricator** (Fig. 4.22) introduces oil mist into the air. The bowl is a reservoir for both the oil and a small quantity of air at slightly higher pressure than the mean supply line pressure. The air passage first restricts the through-flow of air by means of a flexible baffle to give throttling independent of air flow variations. At the point where the air pressure is raised due to the restriction, a passage feeds the bowl via a check valve and applies pressure to the oil which is conveyed to a small reservoir and adjustable needle valve above the air passage. The adjusted oil feed enters a jet placed at the position of minimum air pressure, acting on the venturi (or carburetter) principle.

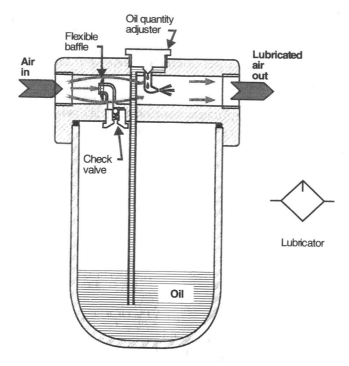

Fig. 4.22 The principle of the lubricator

Exercises for Chapter 4

1. List three treatment processes normally carried out after air has been compressed and before it is distributed.   2.4(b)

2. List three air treatment or pressure control components required in an air supply system at the point of use, after air has been distributed.   2.4(a)

3. For each of the following drying processes, explain as simply as possible, (a) the difference between the methods, and (b) the typical maintainance requirements of the dryer type: (i) refrigerant drying, (ii) absorption drying, (iii) adsorption drying.   2.4(e)

4. Explain (a) why an air distribution main is normally installed out of level; (b) why ring main (loop) systems are often used instead of dead-end mains?
                                                                             2.4(g)

5. List four types of compressor.   2.4(c)

6. List three advantages of multistage compressors as compared with single stage types.   2.4(c)

7. Name the process usually applied between stages of compression in a multistage compressor, and describe briefly the reasons for this feature.   2.4(c)

8. A compressor takes in 500 l/min of free air. If the air delivered by the compressor is converted to a free air quantity (i.e. a calculated quantity in which the compression of the air and change in volume due to heating are converted to the free air equivalent) and the result is an FAD of 400 l/min, what is the volumetric efficiency of this compressor?   2.3(d)

9. List six features in a well designed distribution system which promote trouble-free operation and easy maintenance.   2.4(g)

10. Draw the ISO1219 diagrammatic symbols for:
   (a) relief valve
   (b) pressure regulator
   (c) service unit (or 'FRL')
   (d) lubricator.   2.6(a)

11. Describe two maintenance requirements for a lubricator.   2.4(f)

12. Explain briefly the purpose and characteristics of a soft-start valve.   2.4(f)

13. Briefly describe the purpose of an air receiver.   2.4(a)

14. Draw a diagrammatic freehand circuit of a typical simple compression/distribution system.   2.4(a)

5
Routines

As explained in Chapter 1, the scheme to which this book is mainly devoted has the main purpose of teaching practical skills.

The excellent craftsman, whether in medicine, music or maintenance, builds on the skill of others. The acquisition of skill in pneumatics, the only route which will enable you to deliver results safely and economically, can be progressed by taking on board this simple framework of suggested routines, which are basic summaries of good working practice to get you started.

5.1 Procedures applicable to all types of work on air plant and systems

The following procedures should be adopted:

- Components, tubes, hoses and pipes must be correctly stored, i.e. clean, protected against corrosion, not subject to distortion and clearly identified.
- Up-to-date records and information systems must be maintained as required by regulations and in the interests of efficient maintenance.
- Before planning changes to the maintenance or operation of any item of plant, they should be discussed with all personnel concerned.
- Previous records and manufacturer's documentation should always be consulted.
- A written plan should be prepared for any new operation.
- Availability of components for any repair work anticipated should be investigated before work is commenced.
- All air pressure should be completely isolated and vented from the system together with all electrical supplies, except in the case of 'inspections under working conditions'.
- Ensure no loads remain applied to the system actuators.
- Protect against inadvertent starting of the appliance by 'permit to work' procedure or applying labels to controls etc.
- The correct tools should be available for the job.

- Adequate lifting gear should be available.
- Information arising from the work should be recorded and procedures should be amended to take account of experience.

5.2 Planning installation and commissioning

It is necessary for the following procedures to be planned:

- the order in which installation is undertaken;
- site preparation;
- ensuring component cleanliness, allowing for preparation for assembly, positioning and security;
- siting and connection of compressor and associated equipment;
- tube, hose and pipe installation, associated fittings and assembly procedures – use of various types of fittings;
- pipework installation, layout, fastening and leakage prevention;
- the safety checks associated with initial start-up conditions;
- the start-up procedures, component adjustment and associated settings related to final commissioning;
- the review of current and amended legislation applicable to plant;
- the safety and operational checks associated with a newly commissioned system under load conditions and fitness for use.

5.3 Safety in system design

The following safety measures should be incorporated into any system design:

- All parts of a system should be protected against overpressure.
- Components must have working pressure rating above system pressure.
- Compressed air supply installations must include safety valves, and air receivers must be marked with their design and test pressures.
- Electrical equipment must be properly installed, fused and earthed where required.
- All systems should have a main air-line shut-off valve.
- System design must allow for renewal and adjustment of components.
- Design must be fail-safe.
- Failure must not result in danger to personnel or equipment.
- Scheduled checks must ensure design parameters are never exceeded.

(For an outline of the requirements of the Pressure Systems and Transportable Gas Container Regulations 1989, applicable in addition to the suggestions above, see Ch. 8.)

5.4 Precautions for installation, commissioning and maintenance

The following precautions should be taken:

- Check that no plastic caps or closures remain in components.
- Check pressure rating of all replacement parts fitted.
- Ensure that air connections are mechanically tight and secured.
- Double check that pipe connections are made to the correct ports.
- Ensure that all electrical devices are properly installed, suitably fused, insulated and earthed.
- Ensure that sequence or function of plant is fully understood.
- Check the installation under fully safe conditions, i.e. with guards or other safety arrangements in operation.
- Before commencing maintenance, shut-off air supply and exhaust downstream pressure.
- After repairs or maintenance, ensure any test equipment is removed.
- Always take account of stored energy that may be present in actuators or reservoirs.
- Run system at lowest practicable pressure.
- Ensure plant records are properly completed.
- If checking actuator alignment, whether under pressure or not, ensure that fingers or other parts cannot be trapped.
- Before start-up, ensure all actuators are in correct 'at rest' positions.
- Pressure should be introduced slowly via manual pressure regulator.

5.5 Safety in system operation

When operating the system, be sure to adopt the following safety precautions:

- Never subject the body to compressed air by misapplication of jets of air or by attempting to block exhausting orifices.
- When starting or adjusting a system, take care that fittings do not blow out under pressure.
- Always run a system at the lowest practicable pressure.
- Silencers, exhaust port filters and other types of port fittings should not be removed whilst the system is pressurized.
- Understand the system cycle.

5.6 Maintenance

Maintenance repairs are the response to problems, e.g.:

- excessive noise – which can result from mechanical wear, loose pipework or defective exhaust silencer etc.;

- vibration – resulting from mechanical wear or friction caused by contamination or poor lubrication and also relief valve hammer etc.;
- high temperatures – ineffective cooling or more often a sign of friction caused by contamination or poor lubrication;
- contamination – from poor quality pipe installation, disintegrating seals, water entrainment or non-attention to filters;
- erratic operation – from low air pressure, loose components etc.;
- leakage – poor quality or badly maintained pipework, seal failure;
- excessive pressures – from exceeding the design loading of the system or from friction due to contamination or poor lubrication. Pressure control valves may have been misadjusted to compensate;
- incorrect speed of operation – caused by leaks, or internal slippage due to worn components or seals. Some air tools incorporate governors to limit speed which can malfunction;
- incorrect sequence of operations – resulting from loose or defective proof-of-position valves, tampering with or poor electrical installation of PLCs, loose connections to solenoids etc.;
- hose, tube and pipe failures – caused by vibration, non-allowance for expansion and contraction, movement of loose components etc.

Routine maintenance is compliance with the servicing requirements of equipment at the intervals specified by manufacturers. It is subject to adjustment if necessary to take account of experience and is normally carried out at scheduled periods for the inspection and evacuation, cleaning or replacement of drains, moisture traps, separators, filters etc., and the checking and refilling of lubricators, together with the various service requirements of compressors or other machinery.

Preventative maintenance is commonly embodied in a **plant preventative maintenance (PPM) system**, i.e. a schedule of inspections and good housekeeping practices designed to avoid the occurrence of problems, which is principally based on experience. Preventative maintenance should anticipate (and therefore prevent) problems arising with:

- tools – by planned attention to lubricators, hose condition, mechanical wear, governor operation and local filters;
- machines – by planned inspections under working conditions to detect leakage, high temperatures, pressure changes, loss of speed, mechanical problems or uncharacteristic noise etc.;
- compressor plant – by planned checking of belt tensions if applicable, oil levels, identification of sticking valves, cleaning coolers and checks on cooling water circulation etc.

More detail regarding maintenance is provided in Chapter 18 including the setting up of a planned preventative maintenance system, together with legal requirements and the safety of maintenance personnel.

5.7 Fault-finding

Before carrying out work on a system, personnel should:

- study the circuit diagram (ascertaining that it is an up-to-date version);
- determine the sequence of operation;
- identify the symptoms of the fault;
- list possible causes;
- consider faults other than pneumatic, e.g. mechanical or electrical;
- consider any checks that can be made with the system 'dead';
- identify any safety risks resulting from the failure or the fault-finding procedure.

A preliminary examination should then follow, isolating the system when and where required. The list of possible causes should then be modified accordingly.

A **logical fault-finding procedure** should then be carried out, taking particular note of any hazards; for example, where guards are to be removed or 'proof of position' valves operated, one should be aware of the consequences or associated hazards. After locating the fault:

- remedy the fault
- test the operation
- return the machine to a safe condition.

Consider the cause of the fault: it may be a weakness in the design of the system; if so, make recommendations. A report for the records must then be completed.

A typical logical fault-finding procedure comprises:

- a copy of the circuit diagram with a numbered sequence of test-point positions;
- a list of the test-points with the results noted.

The intention should be to compile data to make fault-finding easier in future. This includes:

- algorithm and functional charts
- step-by-step diagnostic sequences
- fault, cause, remedy (FCR) procedures.

More detail regarding fault-finding is provided in Chapter 19, including notes on common faults, the use of test equipment, functional and algorithm charts, and an explanation of diagnostic methods and fault, cause, remedy procedures.

Exercises for Chapter 5

1. For the construction of a typical small pneumatic system comprising two linear actuators, a motor and solenoid valves etc., supplied from an existing distribution plant, prepare lists of:

(a) items to be prepared or checked before starting installation
(b) items to be checked before commissioning
(c) a commissioning procedure
(d) a safe shutdown sequence
(e) the safe working practices and any regulations applicable to construction and operation of the system.   2.8(a)

2. A regulator is an adjustable valve to limit line pressure. Explain how you would use the regulator when setting up a new circuit?   2.8(h)

3. Typical practical tasks that are set feature the construction of circuits including actuators and control valves. The City & Guilds assessment criteria for marking practical assessments include:

    1.2.2 Start-up and commissioning procedures correctly specified and followed.
    1.2.5 Safe working practice and statutory regulations followed at all times.

    List items suitable for including in the documentation of task procedures: eight items under 1.2.2, and five items under 1.2.5.   2.8(i)

4. Describe briefly five procedures which are normally required when planning a maintenance operation on a pneumatic machine (before the commencement of any practical work) in order to ensure, as far as possible, that the operation is successful.   2.8(b)

5. Summarize six items relating to safety in system design.   2.8(c)

6. Describe briefly what is meant by 'stored energy' which may be present in a machine after shutdown.   2.9(a)
    What are the reasons for 'stored energy' remaining in a system?
    Name two typical causes.

7. Summarize six procedures which should be carried out as necessary preparation before carrying out a fault-finding investigation on a defective machine.   2.9(c)

6
The applications and advantages of pneumatics

6.1 Transmission of energy

The factory compressed air system is, like the electrical installation, a way of supplying energy to a required location. Electrical energy has been generated at the power station for use in the factory or home, and pneumatic energy is generated in the factory for use on site. 'Generation' means conversion either from a natural source such as oil, coal or nuclear materials, or from an energy form resulting from a previous conversion process, as with most compressed air which has been generated with the use of electrical energy.

6.2 Doing work

To convey energy, **work is done** on a medium for transmission to another location where the medium **does work**. The work done in a given time is **power** for which the unit of measurement is the watt or kilowatt (or horsepower in imperial units) as defined in Chapter 2. For our calculations at this level, we use the basic properties of force and speed (distance/time) separately, but when combined they equal power.

To generate electrical energy, work is done on atoms in a conductor to impart a charge. At the place of use, the process is reversed, enabling the electricity to do work. Compressed air is similar but easier to understand: work is done on the air by the compressor, then the actuator or motor does the work at the place of use (See Fig. 6.1).

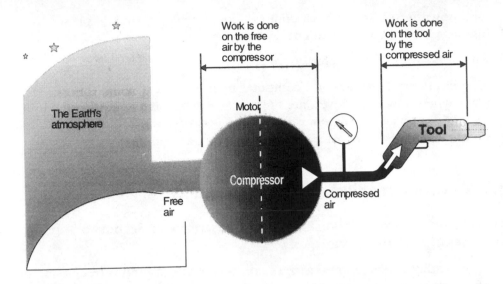

Fig. 6.1 How work is done on, and by, the air

6.3 Scope of pneumatic machines

For any plant where work is done, the available forms of energy transmission are in competition. The designer will choose an energy medium to do the job required, which has minimum first cost, delivers an acceptable quality of product and is most efficient in terms of both running and maintenance costs.

Pneumatics can offer unrivalled versatility because air can be used to:

- drive very fast moving actuators, e.g. for material handling;
- power quite high force actuators, e.g. for extrusion and moulding;
- drive fast and quite powerful motors, e.g. for conveyors;
- drive very light and compact motors, e.g. for tools (see Fig. 6.2);
- drive rotary actuators of all types and sizes;
- sense the presence of materials for control or safety functions;
- provide a lubrication medium for high speed bearings;
- provide a fluid medium for spraying or grit blasting etc.;
- provide a cleaning fluid for removal of swarf or debris;
- provide a vacuum for fluid collection or material lifting applications;
- power plant process control valves which are compact and versatile;
- power equipment in hazardous environments;
- provide a locally stored energy source which is available when the prime energy source has failed.

However, hydraulic fluid may be the preferred energy medium for:

- precise speed control or control of position;
- high force actuators or high-torque motors and rotary actuators;

- high power levels in the most compact equipment, e.g. mobile plant;
- high levels of efficiency and energy conservation.

Electrical power may be preferred for:

- frequent control changes and computer monitoring, e.g. some robots;
- plant where low noise levels are important, e.g. hospital equipment;
- the highest levels of efficiency and energy conservation;
- simple machines with just a few rotational or linear inputs.

6.4 Advantages and disadvantages of pneumatics

In addition to the exceptional versatility described in Section 6.3, pneumatic machines offer further advantages:

- Air is readily available, expensive to compress but the fluid is free.
- After use, air can be vented to the atmosphere.
- Typical factory systems require no return pipework.
- Some system leakage is acceptable.
- Compressed air is easily stored so can provide a reservoir of energy.
- Control of force, speed and direction is achieved with simple devices.
- Overload protection for actuators and motors is a standard feature.
- Components are relatively inexpensive and plant capital cost is low.
- Users are not subject to any health risk due to the fluid itself.
- The generally simple technology is easy and inexpensive to maintain.
- Air actuators and motors feature very fast response to control inputs in comparison with electrical actuators and motors.
- Air equipment is not affected by hot, dusty and wet surroundings.

However, in comparison with hydraulics and electrical equipment, air has some disadvantages:

- The compression, storage and distribution of compressed air are less efficient and so more costly.
- Air actuators are typically six times bigger than hydraulic actuators for the same force output.
- Air actuators cannot deliver very accurately controlled movements.
- Air actuators cannot provide rigid positioning because of air's compressibility.
- The energy which can remain stored in air systems when they are shut down can be very dangerous.
- In recognition of the dangers of stored air, comprehensive safety regulations apply.
- Air compressors, motors and tools can be noisy but have improved in this respect.

Fig. 6.2 Two 0.25 kW motors – a 450 rpm air motor weighing 1 kg shown beside a 1725 rpm electric motor which weighs 7 kg. (Photograph by kind permission of Ingersoll-Rand Corporation.)

Exercises for Chapter 6

1. Name three industrial applications known to you which could be achieved by air, hydraulics or electrical sources of energy, but for which air has been preferred. State the disadvantages of hydraulics and electrics which have contributed to the choice of pneumatics in each case. 2.2

2. List 10 ways in which air can be applied in the workplace. 2.1

3. List two industrial applications for which air is the only applicable process. 2.2

4. List two industrial fluid power applications for which air is probably unsuitable, and for which hydraulics is the preferred choice. 2.2

5. List two industrial fluid power applications for which air is probably unsuitable and for which electrical devices are the preferred choice. 2.2

6. List two industrial power applications for which air is probably unsuitable and for which purely mechanical devices are the preferred choice. 2.2

7. Describe briefly the purpose of pneumatic actuation as the means of providing linear/rotational power and movement and give six examples of this. 2.1

7
Properties and behaviour of air (2)

7.1 Humidity

Water in air destroys machinery and degrades its performance, so, as shown in Chapter 4, the air is treated to remove most of the water. In order to specify an air dryer and plan the maintenance necessary to support the drying regime, the plant designer needs to know how much water must be removed. Tests are therefore carried out to provide data for calculations giving the water content of the air under working conditions.

All air contains some water, but the water is invisible unless there is so much of it that mist, or droplets of water, become suspended in the air and deposited on surrounding surfaces. As the moisture content of a volume of air is increased, the formation of mist in the air occurs at a definite measurable point, called the **dew point**. When it reaches the dew point, the air is **saturated**. The pressure and temperature of the air also affect the dew point – as pressure rises air can hold less water, and as temperature rises air can hold more water. This is demonstrated in a car on winter mornings – at cold surfaces such as windows the dew point falls, so mist and droplets form, a problem which is cured by heating the window.

Typically, the air around us does not contain visible mist and so is not saturated. The amount of water in a volume of air, in comparison with the amount it could hold at the dew point of the air, is called the **relative humidity**. The typical relative humidity of air varies considerably from one locality to another, and it must therefore be tested (using a 'hygrometer') before the calculations can be made to enable an air treatment plant to be specified.

Definition of relative humidity

Relative humidity is the actual quantity of water in air, divided by the maximum quantity which the air can contain at the same temperature and pressure, expressed as a percentage

---

If the relative humidity of a sample of air is known, it can be converted into a quantity of water at any required temperature and pressure using the information in Fig. 7.1.

Figure.7.1 is based on 1 cubic metre (m³) or 1000 litre of saturated air (i.e. air which has reached its dew point) at an atmospheric pressure of 1 bar and a 'standard temperature' of 15°C. This sample of saturated air will contain 12.83 gram (g) of water. The graph simply shows the effect of pressure or temperature changes, or combinations of both, on the maximum quantity of water which the sample can hold. If 1 m³ of air is raised to 7 bar and cooled to 20°C, then its water holding capability is reduced to only about 2.2 g, so 12.8 − 2.2 = 10.6 g will have dropped out as droplets or sediment, or will alternatively have been safely removed by the dryer.

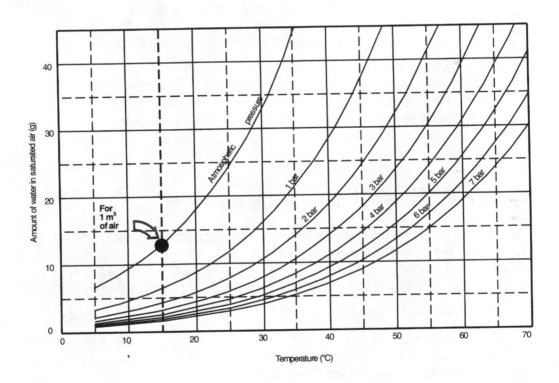

Fig. 7.1 The quantity of water in saturated air

Worked example 7.1

Calculate the quantity of water deposited (in g) when 1 m³ of free air with a relative humidity of 90% is compressed to 6.0 bar with a final temperature of 45°C.

(a)   Calculate the water in the 1 m³ of free air. Free air with 100% relative humidity has 12.8 g of water per m³, so with 90% relative humidity it will hold 12.8 × 90% (= 12.8 × 0.9) = 11.5 g/m³

(b)   Read from Fig. 7.1 the quantity of water in saturated air at 6 bar and 45°C. From the figure, air at 6 bar and 45°C can hold 10.5 g/m³

(c)   Subtract the maximum water the compressed air can hold (b) from the water held in the free air (a).

The water deposited will be 11.5 − 10.5 g = 1.0 g Ans

This is a typical relative humidity calculation, but assessment questions are often simpler. Note that relative humidity can be quoted as a percentage, e.g. 90%, or as a decimal fraction, e.g. 0.9, both with the same meaning.

7.2 Effect of changes in the air or system on pressure

Figure.7.2 shows a compressed air supply powering a loaded actuator; all the components in between have been removed so that the basic facts are as clear as possible. If an air supply at the point of use is connected to a simple system to do work, the pipeline pressure is created by the load. In the system shown in Fig. 7.2,

$$\text{pressure } P_1 = \text{load/piston area}$$

The small pressure drops due to the friction of air in the pipe are represented by p_1 (roughness), p_2 (bends) and p_3 (change of section), pressure differences created due to the movement of air past these obstructions.

   If the air pressure is holding up the load but not moving it, then there is no movement of air and hence no friction, and pressures p_1, p_2 and p_3 are all zero. The pressure P_2 where the air enters the system is then equal to load pressure P_1. But when air starts to move to raise the load, then p_1, p_2 and p_3 all become significant pressure drops, so

$$\text{pressure } P_2 = P_1 + p_1 + p_2 + p_3$$

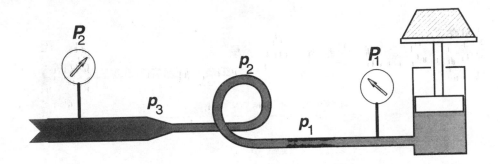

Fig. 7.2 How pressure changes through a pipeline

Therefore, when air is moving in a typical system doing work, the pressure is higher where the air enters than at the point where the load is applied. The inlet pressure is the load pressure plus pipeline and valve losses. For a given pressure and flow, changes in temperature will not affect the pressure.

7.3 Effect of changes in the air or system on flow

If air quantity is corrected to free air, flow is simple. What goes in must also come out, so flow out equals flow in minus any leaks, as shown in Fig. 7.3.

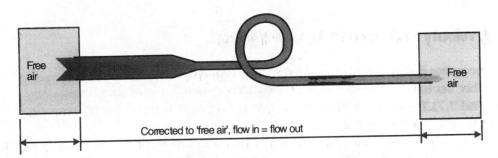

Corrected to 'free air', flow in = flow out

Fig. 7.3 Flow in terms of free air

Since actual air volume is much less when compressed, the real flow in a machine or component must often be considered; flow in terms of free air is not appropriate. The effects of pressure and temperature change then become very complex and the results sometimes surprising; for example, the flow into a pipeline is often less than the flow out. The subject is best considered by two hypothetical situations: that shown in Fig. 7.4 in which temperature changes are ignored, and that in Fig. 7.5 in which pressure changes are ignored.

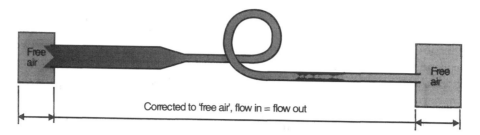

Fig. 7.4 Flow in < flow out

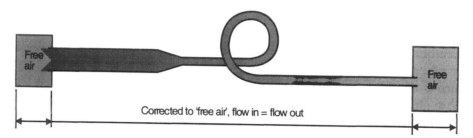

Fig. 7.5 Flow in > flow out

However, typically flow varies due to both pressure change and cooling, but the effect of pressure is much greater. A temperature loss of 10°C results in only 3.5% less volume, so the effect of temperature can often be ignored.

7.4 Velocity and pressure drop in pipework

Nomograms for air velocity and pressure drop in pipes provide quick, easy and reasonably accurate answers as an alternative to troublesome calculations (see Figs 7.6 and 7.7). They also illustrate how variables (such as pressure) influence the size of pipework necessary.

When a new system is planned or an old one extended, it is essential to size the pipework to handle present and anticipated flow demand. The information required is the estimated length of pipework together with the total flow in terms of free air and the normal minimum working pressure – minimum since it is at minimum pressure that the actual flow will be greatest.

Pipes are sized using two criteria: velocity and pressure drop. The size is first chosen to give a reasonable velocity, within a range indicated by experience, and then checked for an acceptable pressure drop over the total length.

Velocity can be calculated with a simple formula, or more easily using Fig. 7.6. The pressure drop for pipe can be calculated with a rather complex formula or by using Fig. 7.7. The pressure drop through fittings and components, worked out with manufacturer's data, is then added.

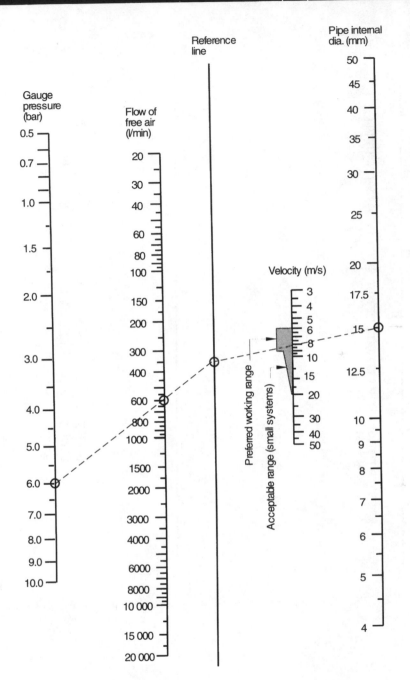

Fig. 7.6 Nomogram for obtaining air velocity in pipes.
How to use a nomogram. Circle three known values. Draw a line (see example) through two of them extended to the reference line. A line then drawn from the intersection with the reference line to the third known value will enable the unknown value to be read from the scale.

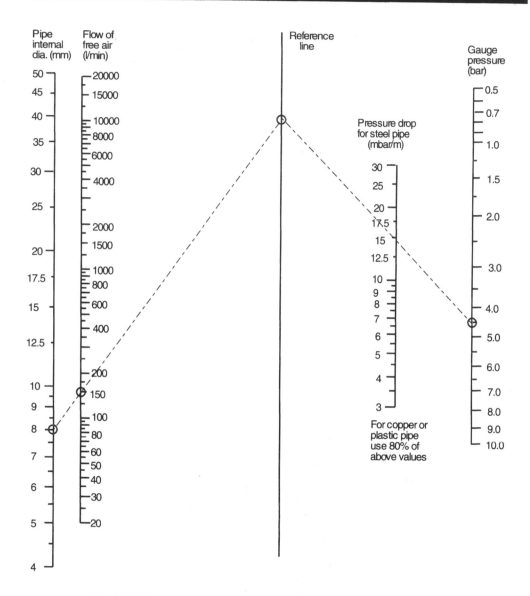

Fig. 7.7 Nomogram for obtaining pressure drop through pipes

Exercises for Chapter 7

1. Calculate using the nomograms (Figs 7.6 and 7.7) the smallest of the following standard medium weight steel pipe bore internal diameters that will pass an FAD of 400 l/min at compression of 6 bar, (a) without exceeding max. pressure loss of 8.0 mbar/metre of length, (b) without exceeding a maximum air velocity of 6.0 m/s.

Nominal bore (inches)	Actual internal diameter (mm)
1	27.6
3/4	21.7
1/2	16.1
3/8	12.8
1/4	8.8

Note: two separate answers (a) and (b) are required to this question.

2.3(i)

2. Use the nomograms (Figs 7.6 and 7.7) to calculate:
   (a) The pressure drop (bar) with an FAD of 7500 l/min at 7.0 bar average pressure through 20 m of 40 mm i.d. steel tubing.
   (b) The FAD (in l/min) which gives an air velocity of 7.0 m/s at an average pressure of 7 bar through a 12.5 mm i.d. pipeline.   2.3(i)

3. Define 'relative humidity'.   2.3(g)

4. During the distribution of air after compression, it is subject to changes. State the effect of each of the following on the air flow at the change point (i.e. on the quantity of air as compressed, not as FAD): (a) the air cools; (b) obstructions and constrictions in the pipeline; (c) the pressure is reduced; (d) the pipe diameter is reduced.   2.3(h)

5. For an FAD of 800 l/min, compressed to a pressure of 7.0 bar, travelling through a steel pipe with an internal diameter of 15.0 mm, use the pipe sizing nomograms (Figs 7.6 and 7.7) to calculate:
   (a) The air velocity in m/s that will be reached.
   (b) How much pressure would be lost (in bar) if the 15 mm bore pipe was a straight length of 50 m?
   Is the air velocity reached in (a) above:
      (i) normal for a well designed distribution system?
      (ii) high but within acceptable limits for a small system?
      (iii) too high and therefore unacceptable?   2.3(i)

6. This small cylinder has a full bore piston area of 700 mm² and a rod-end piston area of 600 mm². To restrict speed the flow restrictor has been adjusted until the incoming (left-hand) pressure gauge reads 6 bar. Neglecting losses, what will be the reading of the right-hand gauge?   2.3(a)

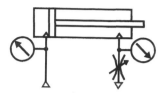

7. Suppose the cylinder in question 6 above is supplied via steel delivery pipework without bends, 100 m in length with internal diameter of 10 mm, and that the quantity of flow to the cylinder is set at an FAD of 200 l/min. Calculate using the nomogram (Fig. 7.7) the approximate resulting supply pressure at the delivery pipe inlet.   2.3(i)

8
Legal requirements for air systems

8.1 The Pressure Systems and Transportable Gas Regulations 1989

Compressed air can be extremely dangerous. A small air receiver just 500 mm in diameter and 1 m long containing compressed air at 7 bar pressure is resisting an internal force totalling 135 tonne. For a receiver twice the size, the figure would be 500 tonne. Before legally binding regulations were introduced, compressed air systems were the major cause of serious accidents at work.

Due to the application and enforcement of 'The Pressure Systems and Transportable Gas Regulations', together with applicable British Standards and other related Codes of Practice, accidents with compressed air are now unusual, so much so that the dangers may be forgotten.

Figure 8.1 is a chart showing when the pressure systems regulations are applicable and giving an outline description of the formal written scheme which the owner or user of the system is responsible for establishing under the regulations. A 'competent person' must be appointed to advise the user on the scope of the written scheme, to draw up and certify schemes of examination and carry out examinations under the scheme. The user is responsible for appointing a competent person who has the necessary knowledge and experience and who has a position enabling independent advice to be given, the level of qualification depending on whether the system is of minor, intermediate or major status, as defined in the rules.

For the study level covered by this book, only outline knowledge regarding the pressure systems regulations is required. A useful list of summarized headings with brief descriptions is as follows:

1. System operating limits must be marked and advised.
2. There must be a written scheme for examination and inspection.
3. A competent person must be appointed to give advice, prepare the scheme and carry out the examinations.
4. Operating instructions must be available, i.e. manufacturers and installers data for the plant.
5. A maintenance plan must be written and followed.
6. Records of inspections and any modifications to the plant must be maintained.

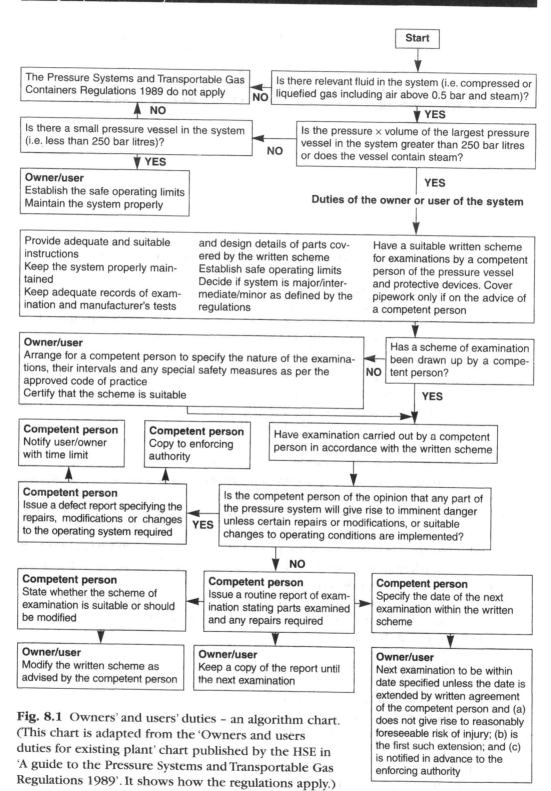

Start

Is there relevant fluid in the system (i.e. compressed or liquefied gas including air above 0.5 bar and steam)?

NO → The Pressure Systems and Transportable Gas Containers Regulations 1989 do not apply

YES ↓

Is the pressure × volume of the largest pressure vessel in the system greater than 250 bar litres or does the vessel contain steam?

NO → Is there a small pressure vessel in the system (i.e. less than 250 bar litres)?

YES ↓

Owner/user
Establish the safe operating limits
Maintain the system properly

YES

Duties of the owner or user of the system

Provide adequate and suitable instructions
Keep the system properly maintained
Keep adequate records of examination and manufacturer's tests

and design details of parts covered by the written scheme
Establish safe operating limits
Decide if system is major/intermediate/minor as defined by the regulations

Have a suitable written scheme for examinations by a competent person of the pressure vessel and protective devices. Cover pipework only if on the advice of a competent person

Owner/user
Arrange for a competent person to specify the nature of the examinations, their intervals and any special safety measures as per the approved code of practice
Certify that the scheme is suitable

NO ← Has a scheme of examination been drawn up by a competent person?

YES

Competent person
Notify user/owner with time limit

Competent person
Copy to enforcing authority

Have examination carried out by a competent person in accordance with the written scheme

Competent person
Issue a defect report specifying the repairs, modifications or changes to the operating system required

YES ← Is the competent person of the opinion that any part of the pressure system will give rise to imminent danger unless certain repairs or modifications, or suitable changes to operating conditions are implemented?

NO ↓

Competent person
State whether the scheme of examination is suitable or should be modified

Competent person
Issue a routine report of examination stating parts examined and any repairs required

Competent person
Specify the date of the next examination within the written scheme

Owner/user
Modify the written scheme as advised by the competent person

Owner/user
Keep a copy of the report until the next examination

Owner/user
Next examination to be within date specified unless the date is extended by written agreement of the competent person and (a) does not give rise to reasonably foreseeable risk of injury; (b) is the first such extension; and (c) is notified in advance to the enforcing authority

Fig. 8.1 Owners' and users' duties – an algorithm chart. (This chart is adapted from the 'Owners and users duties for existing plant' chart published by the HSE in 'A guide to the Pressure Systems and Transportable Gas Regulations 1989'. It shows how the regulations apply.)

8.2 New legislation – the CE mark

European safety legislation recently introduced includes:

- the **pressure vessels directive** applicable to series manufactured air receivers;
- the **machinery directive** applicable to all machines with moving parts, including lifting equipment and safety components.

Manufacturers and suppliers are responsible for ensuring that their products when sold comply with the standards and approvals which the mark signifies.

The pressure vessels directive applies to all air receivers of welded construction for more than 0.5 bar pressure, whereas the pressure systems regulations apply only from about 25 litre capacity upwards. They must be manufactured using sound engineering practice and be marked with their pressure rating, capacity, temperature range and serial number. Receivers over about 5 litre capacity, which must now have the CE marking, are required to have a safety examination and certificate from an approved body.

The machinery directive requires machine manufacturers to carry out a risk assessment to allocate levels of risk to the various functions of a machine, and to draw up a technical document which explains the safety measures incorporated, in compliance with published standards and requirements. All assemblies of linked components with moving parts must comply and carry the CE mark.

8.3 New legislation – condensate disposal

All lubricated compressors with aftercooling must now fit a device to remove traces of oil from the discharge of cooling water. Certain specially formulated lubricants are available for some compressors which ensure that any trace in the condensate is biodegradable into harmless substances and may therefore be discharged without precautions.

Exercises for Chapter 8

1. Explain briefly why compressed air is dangerous.   2.8(i)

2. Give six principal headings which summarize the Pressure Systems and Transportable Gas Regulations 1989, and for each heading give a very brief explanation of the requirements.   2.4(d)

3. Is distribution and machinery pipework covered by the legal requirements of the Pressure Systems and Transportable Gas Regulations 1989?   2.4(d)

4. Outline the legal requirements arising if, in the course of a periodic examination, the competent person is of the opinion that any part of the system will give rise to 'imminent danger'

2.4(d)

5. The Pressure Systems and Transportable Gas Regulations 1989 apply (except in the case of steam) to systems with pressure vessels of more than 250 bar litre capacity [i.e. pressure (bar) × volume (litre) > 250]. Does the recent European legislation result in a requirement for smaller pressure vessels to be examined and certified?

2.4(d)

Note

Readers and users of *Practical Pneumatics* should note that the summarized information above regarding the Regulations reflects the level of knowledge required from candidates for the City & Guilds 2340 Part 2 schemes in Fluid Power Engineering. This information should not be regarded as an effective summary of the legal requirements for other purposes or a substitute for the regulations and explanatory publications available from the Health and Safety Commission.

9
Pneumatic components (1)

9.1 Linear actuators

The purpose of any pneumatic machine is to perform work, in most cases by actuators, either linear actuators, rotary actuators or motors.

The linear actuator or cylinder converts the energy conveyed by the air into straight-line movement and force. The simplest type, the **single-acting cylinder**, is powered by the air in one direction only, and relies on the load or a spring for the return stroke.

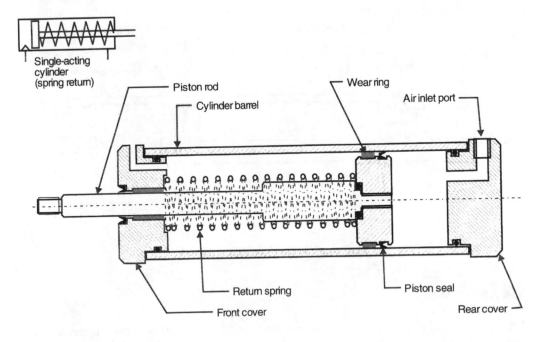

Fig. 9.1 Single-acting cylinder

Cylinder piston seals are of many alternative designs. Figure 9.1 shows a common type used in both single-acting (with one seal) or double-acting cylinders (with two seals) in which a separate wear ring is fitted.

The **double-acting cylinder** is powered by the air in both directions. Figure 9.2 shows the common type of piston seal arrangement similar to Fig. 9.1 and also a fairly common one-piece design.

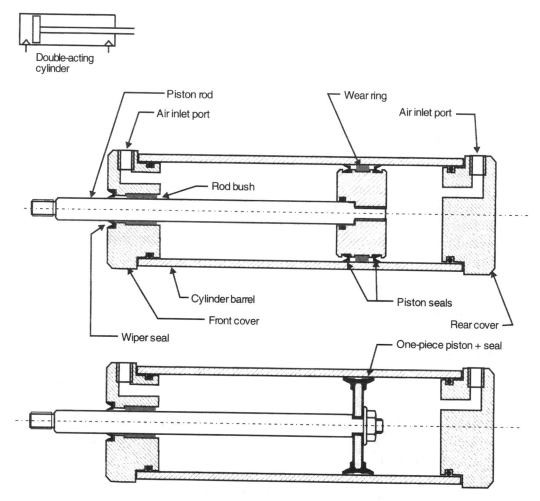

Fig. 9.2 Double-acting cylinders

9.2 Cushioning

Usually it is not acceptable for linear actuators to finish their working stroke at full speed, due to the resulting impact, noise and wear, so **cushioning** is a common added feature. As the piston nears the end of stroke, a projecting sleeve blocks the normal air path so the air must exit through an alternative restricted path. The restriction is often in the form of an externally adjustable needle valve. Other designs incorporate the restricted air path within the cushioning sleeve with no external adjustment.

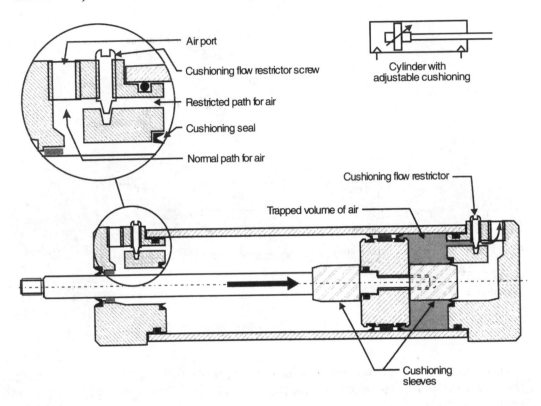

Fig. 9.3 Adjustable cushioning

If cushioning is employed to reduce the speed of heavy, fast-moving loads, it can result in very high pressure within the cylinder, so adjustment should be carried out with caution.

9.3 Calculation of linear actuator forces

From Chapter 3, to calculate force exerted by pressure acting on an area:

force = pressure × area

The units (Ch. 2) are as follows:

- Force – newton (N)
- Area – mm^2
- Pressure – bar

To apply the relationships above, use bar/10 for pressure, so

$$\text{force (N)} = \frac{\text{pressure (bar)}}{10} \times \text{area (mm}^2)$$

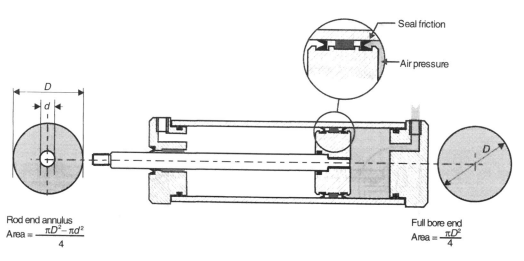

Fig. 9.4 Force developed by a cylinder

The effective piston areas in a cylinder are shown in Figure 9.4. Substituting these for the word 'area' in the equation gives, for the full bore end of the cylinder (extension stroke):

$$\text{force (N)} = \frac{\text{pressure (bar)}}{10} \times \frac{\pi D^2}{4}$$

and for the rod end of the cylinder (retraction stroke):

$$\text{force (N)} = \frac{\text{pressure (bar)}}{10} \times \frac{(\pi D^2 - \pi d^2)}{4}$$

Worked example 9.1

Calculate the extension force (neglecting frictional losses) resulting from a pressure of 6 bar acting on a piston of 40 mm dia.

Force (N) = pressure (bar)/10 $\times \pi D^2/4$ (mm^2)
so force = 6/10 $\times \pi \times 40^2/4 = 0.6 \times \pi \times 1600/4$
= 754 N Ans

Although the 30% allowance for frictional losses is usually specifically excluded in assessment questions, an allowance should always be included in making calculations for real applications.

Worked example 9.2

Calculate the retraction force (neglecting frictional losses) resulting from a pressure of 6 bar acting on a piston of 40 mm dia. with piston rod of 18 mm dia.

Force (N) = pressure (bar)/10 $\times$ ($\pi D^2 - \pi d^2$)/4 (mm^2)
so force = 6/10 $\times$ π $\times$ ($40^2 - 18^2$)/4 = 0.6 $\times$ π $\times$ (1600 − 324)/4 = 0.6 $\times$ π $\times$ 319 = 601 N Ans

---

9.4 Air motors

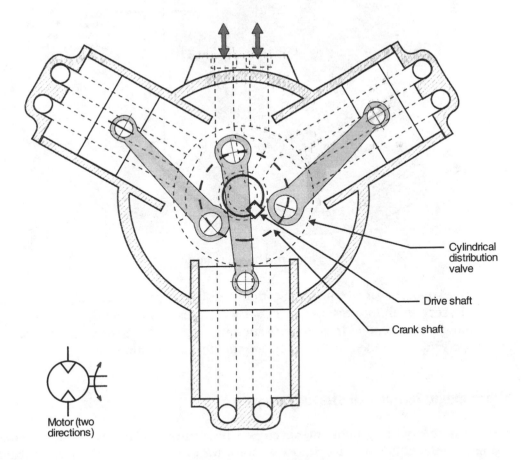

Fig. 9.5 The piston motor

Whereas a displacement type compressor applies force to a contained volume of air to generate pressure, the motor uses a contained volume of air under pressure to create force. This is equally true of the linear actuator, but in a motor the actuators are arranged to rotate the driving shaft. This principle is most easily seen in the piston motor: a cylindrical distribution valve coupled to the crankshaft directs the incoming pressure and outgoing exhaust to and from each cylinder in turn (Fig. 9.5). Piston motors are bulky and efficient, so they are used mainly for slow speed, high torque applications like conveyor drives and winches.

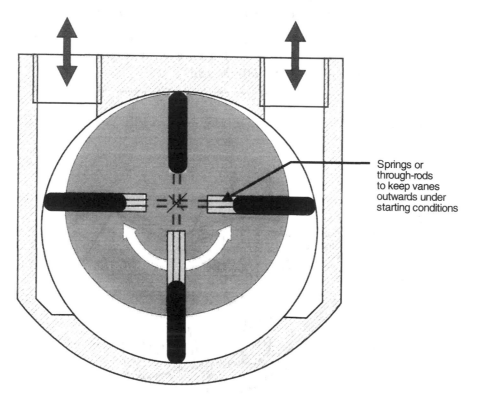

Springs or
through-rods
to keep vanes
outwards under
starting conditions

Fig. 9.6 The vane motor

Many air motors for factory applications are vane motors (Fig. 9.6), which can be made very small for portable tools but which tend to be inefficient. Similar to vane pumps, they tend to have fewer vanes and must include some mechanism to keep the vanes extended so that they are effective at start-up.

9.5 Running torque and starting torque

Torque is the twisting moment developed by a motor or rotary actuator. When using manufacturer's data to choose a motor for an application, the most reliable method is to select a motor which can deliver enough torque at the projected working pressure, and can also achieve the required speed. Horsepower can also be used as a criterion for selection but requires more care in interpretation.

Normally two types of torque are quoted: running torque and static or starting torque. The running torque of a motor is always considerably higher than the starting torque, and the less efficient the motor, the lower will be the relative starting torque. The starting torque of some small vane motors with three or four vanes is dependent on the position of the vanes and can even be zero; in other words manual encouragement is required to get the motor going. As with the calculation of theoretical force for linear actuators, a theoretical capability can be calculated for motors, to which allowances must be applied for application. A guide to the torques available from an efficient piston motor is:

- running torque is about 60% of the theoretical torque;
- starting torque is about 75% of the running torque.

Manufacturer's data should always be used if available.

9.6 Directional control

The term directional control valve refers to all valves which use a mechanical force, air signal or electrical signal to block or divert the direction of an air flow.

The standard ISO1219-1 diagramatic representation of a valve clearly shows the valve function. The valve is represented by a rectangle or box in which are drawn arrows or lines showing the flow. If the box has a single flow path as in Fig. 9.7, then the valve has two ports – in and out. The simplest valve, like a switch, will have two working positions, open and closed, so the box representing the valve is shown twice, each representation showing the applicable flow path.

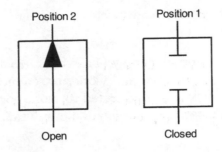

Fig. 9.7 Basic diagram of valve positions

To complete the diagrammatic symbol of the valve, the two boxes are drawn adjacent to each other, and the valve **actuators** are added. An actuator is shown joined to the box with the flow path which takes effect when the actuator is used. A common on–off valve type, for example, is a press-button, spring return, as shown in Fig. 9.8. The press button is one actuator, to make the valve open. The spring return is the other actuator which makes the valve close. The valve has two ports (each box has an 'in' and an 'out') and two positions of use, so is known as a **2/2 valve**. The normal position of the valve (when it is not actuated), also known as the 'at rest' position, has the port positions extended to show the connection points to the valve.

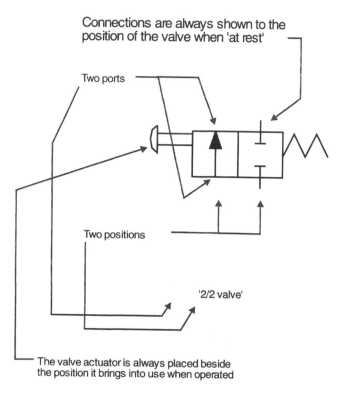

Fig. 9.8 Valve symbols – the basic rules

3/2 valves have three ports – supply (1), exhaust (3) and the work port (2) – and two positions – supply blocked and work port connected to exhaust (so the valve is off or 'closed'), or supply connected to work port (so the valve is on or 'open'). Chapter 10 explains the representation of valves further and how they are applied in circuits.

9.7 Valve mechanisms

Valves are of two main types: the **poppet valve** (Fig. 9.9) which seals on the face or faces of the internal moving part, and the **spool valve** (Fig. 9.10) which seals on the outside diameter of the moving sleeve called the spool. The majority of directional control valves, including all the larger sizes, are spool valves.

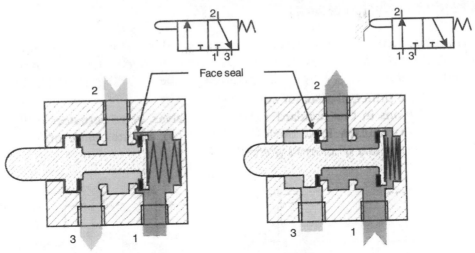

Fig. 9.9 Typical poppet valve

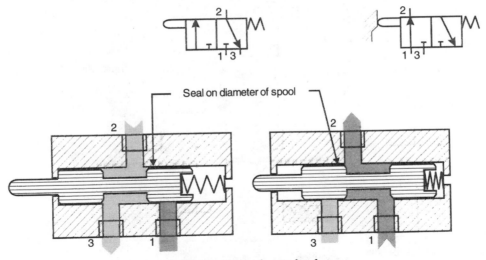

Fig. 9.10 Typical spool valve

Poppet valves are mainly the smaller valves because the spool cannot be balanced. Note that for the spool valve shown in Fig. 9.10, every force acting on the spool is balanced by an equal and opposite force and that both ends of the spool are vented to atmosphere. Because the spool is balanced, the force required to operate the valve is the spring force plus any friction. The operating force is therefore low. Note, on the other hand, that for the poppet valve shown in Fig. 9.9, to depress the plunger in the left-hand view it is necessary to overcome the supply air pressure acting on the full seal diameter plus the spring force. The operating force is therefore high. This inherent design problem with poppet valves limits their application to small flows requiring small internal passages and seals. The disadvantages of spool valves are greater susceptibility to dirt and more resistance to air flow.

9.8 Methods of valve actuation

'Actuation' includes operating levers, air-operated pilot pistons and solenoids etc. and also return mechanisms such as springs or 'detent' mechanisms which provide notched locations for each working position of the valve. Figure 9.11 shows a selection of the principal actuator symbols.

Some variations in the 'standard' symbols for valves, valve actuators and other components, particularly logic devices (see Ch. 14), are still quite often found, depending on the manufacturer's preferences.

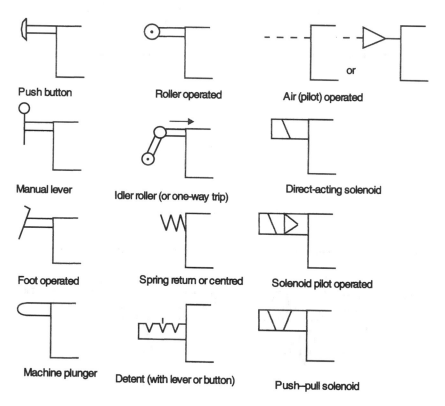

Fig. 9.11 Principal valve actuator symbols to ISO1219-1

9.9 Restriction of flow

The flow restrictor is a needle valve adjusted to obstruct the air path by imposing a pressure drop. The pressure of the air on the supply side of the valve (the 'upstream' air) is raised and, depending on the adjustment of the restrictor, less air passes through to the downstream system. The flow restrictor is therefore an effective speed control for actuators or motors. Figure 9.12 shows the two common types, with or without a non-return valve to give free flow in one direction. The application of flow restrictors is discussed in Chapter 10.

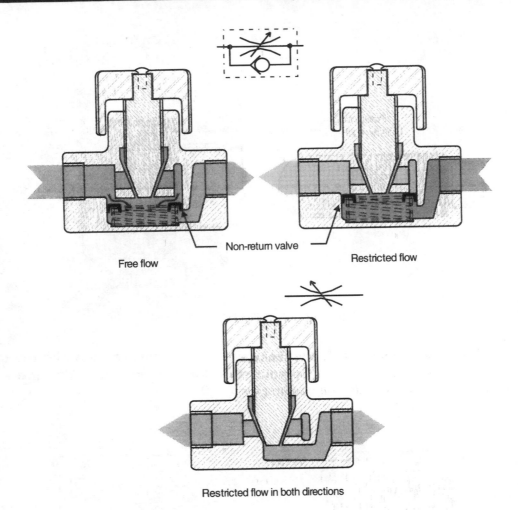

Free flow

Non-return valve

Restricted flow

Restricted flow in both directions

Fig. 9.12 Flow restrictors

Exhaust port flow restrictors are also an effective and economical form of speed control, combining the adjustable restrictor with the function of an exhaust noise silencer, as shown in Fig. 9.13.

9.10 Silencers and reclassifiers

When air is vented to atmosphere, the pressure drop as it is discharged, together with all the pressure disturbances imposed on the air during the process of doing work, pass into the atmosphere to create noise in the workplace. Silencers, typically as shown in Fig. 9.13, are fitted on the exhaust ports of valves, motors and actuators to reduce this noise.

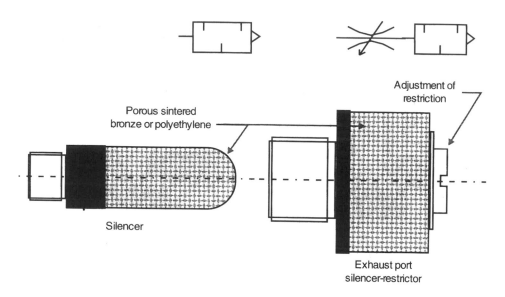

Fig. 9.13 Exhaust port silencers

Reclassifiers or **exhaust filter silencers** are used in the same way as exhaust port silencers, for the additional purpose of removing oil mist from the air discharged for reasons of health and safety (see Fig. 9.14).

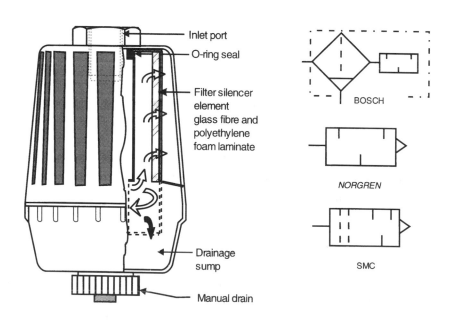

Fig. 9.14 The reclassifier

9.11 Pipes and fittings

Four principal piping systems are used, each suited to different functions in the factory environment from compression through to the application of air (Fig. 9.15):

- Galvanized 'malleable iron' – used for compression and distribution.
- Copper pipe with brass compression fittings – used for local distribution and heavy-duty machine connections.
- Flexible plastic tubing and push-in fittings – light duty machine connections, colour coded for easier troubleshooting.
- Lined ABS piping with solvent-cement joints – distribution in protected environment subject to temperature limits and adequate support.

Note that the imperial 'nominal bore' inch sizes are normal for iron or steel distribution pipework but that other piping systems are sized by the pipe outside diameter in millimetres or inches.

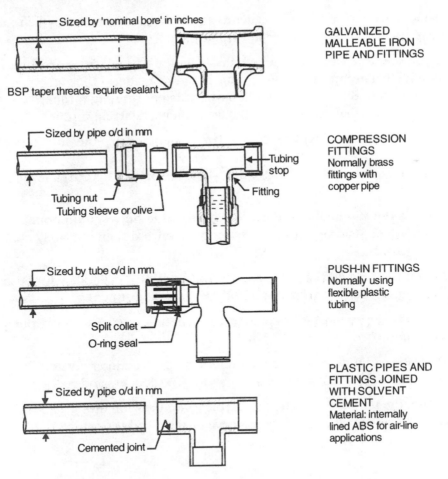

Fig. 9.15 Types of pipes and fittings

Exercises for Chapter 9

1. Describe with the help of a sketch, cushioning in an air cylinder.
   What is the purpose of cushioning?
   Draw the symbol for a double-acting cylinder with cushioning.   2.5(a)

2. Outline two essential differences in construction between (a) a single-acting spring-return air cylinder, and (b) a double-acting air cylinder.
   Draw the ISO1219 diagrammatic symbols for the two actuator types.   2.5(a)

3. For a double-acting cylinder with 80 mm bore dia. and 22 mm rod dia. working at a supply pressure of 4 bar, calculate:
   (a) the actuation force in newton when extending;
   (b) the actuation force in newton when retracting.
   Note: accuracy to the nearest whole number is sufficient. Do not include any allowance for friction losses. Show all your working.   2.5(a)

4. Name one disadvantage of smaller rotary vane type air motors which can make them totally unsuited for certain applications.   2.5(a)

5. If you are asked to recommend an air motor to replace an existing electric motor, what are the allowances that should be made:
   (a) to ensure that the motor will have sufficient running torque;
   (b) to ensure that the motor will have sufficient starting torque?   2.5(g)

6. In describing pneumatic valves, what is signified by the numbers (a) 2/2, (b) 3/2?
   Draw the symbols for: (c) a 2/2 press-button/spring valve; (d) a lever operated 2/2 valve.   2.5(b)

7. What is the essential difference in construction between a poppet valve and spool valve? State one reason why poppet valve design is mainly confined to small air valves.   2.5(b)

8. Sketch the rear end-cap (i.e. not the rod end) of a cylinder with cushioning, showing the functional features and means of adjustment.   2.5(a)

9. What is the principal difference in construction between a vane pump and a vane motor?   2.5(a)

10. Describe in a few words the differences in construction between: (a) silencers and exhaust port silencer/restrictors; and (b) silencers and reclassifiers. 2.5(b)

11. Sketch the tubing connection to a typical compression fitting, indicating the fitting, the tubing stop, the tubing sleeve or olive and the tubing nut.   2.5(e)

10
Pneumatic circuits (1)

10.1 Linking components in diagrams

In addition to the few simple rules governing the creation of valve symbols described in Chapter 9, components are linked in circuit diagrams using the following rules:

- Pressure and exhaust pipelines, and also pipelines serving actuators, are represented by straight lines with right-angle corners.
- Control signal and pilot lines are similarly represented with dotted lines.
- Where pipelines are joined, this is indicated by a small circle or block at the intersection.
- Where pipelines cross without joining, they are simply drawn over each other without the indication of a joint. Bypass loops are not used.
- Components are not always shown in their relative working positions. Positions on the circuit diagram are in the interests of a clear circuit and not to show mechanical relationships.

Figure 10.1 shows the basic symbols for air supply and exhaust connections. Although in an actual circuit there would be one air supply piped to all the components requiring mains air, diagrams frequently show individual air supply symbols attached to each component. This results in a simpler and clearer circuit.

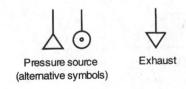

Pressure source
(alternative symbols)

Exhaust

Fig. 10.1 Air supply and exhaust symbols

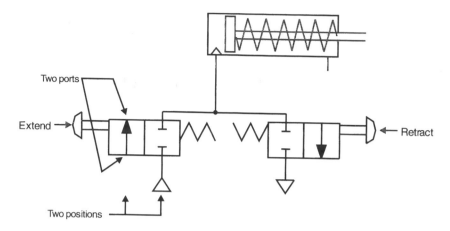

Fig. 10.2 Simple directional control using 2/2 valves

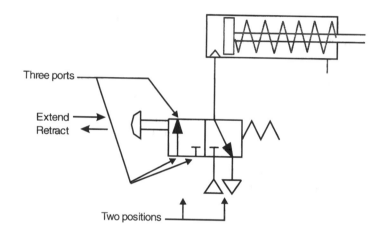

Fig. 10.3 Directional control using a 3/2 valve

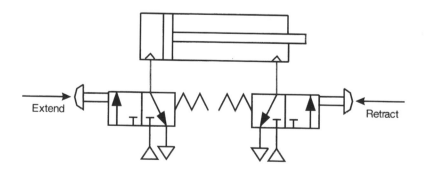

Fig. 10.4 Simple directional control using 3/2 valves

Figures 10.2, 10.3 and 10.4 show directional control of single-acting and double-acting cylinders using 2/2 or 3/2 press-button/spring return valves. Figure 10.5 is more representative of typical practice, in which the actuator is linked with a nearby directional control valve (the circuit does not of course show whether components are close together or separated by miles of pipework) which in turn is pilot operated by a remotely situated press-button spring-return manual valve.

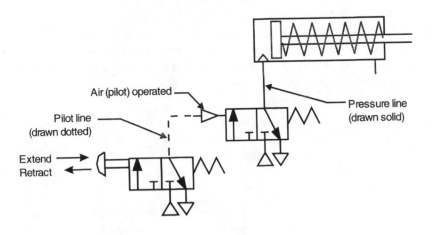

Fig. 10.5 Typical piloted directional control

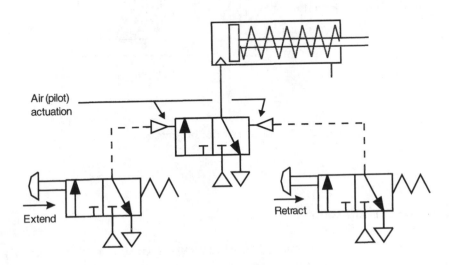

Fig. 10.6 Piloted control of single-acting actuator

The spring-return valves in the circuits in Figs 10.5 and 10.7, whether air operated or direct-acting, require continuous actuation to deliver a supply to the actuator. The directional control valve used in typical applications has no spring return but is of double-air operation or pilot–pilot type as in Fig. 10.6. A short duration pilot signal is sufficient to change the valve position, which remains as selected, delivering air continuously to the actuator until the opposite pilot is operated. The ability of a two-position pilot–pilot valve to 'memorize' its working position is an essential feature of most practical pneumatic circuits.

A double-acting actuator, powered in both directions, has two ports and so must be supplied by a directional control valve with two work ports. 4/2 and 5/2 (four or five ports and two positions) valves have the same function, but a 5/2 valve has two exhaust ports to enable separately adjusted exhaust port flow restrictors to be fitted, for example, and is the more common type. Figure 10.7 is a simple circuit for a 5/3 pilot spring-return valve.

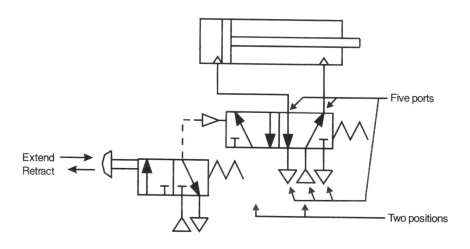

Fig. 10.7 Control of double-acting actuator

10.2 Reciprocating control

The 'limit-switch' valve is a small 3/2 spring-return type actuated by a roller plunger for the purpose of mechanically monitoring an actuator movement and initiating the following movement in the sequence. The simplest application of a limit switch valve is shown in Fig. 10.8.

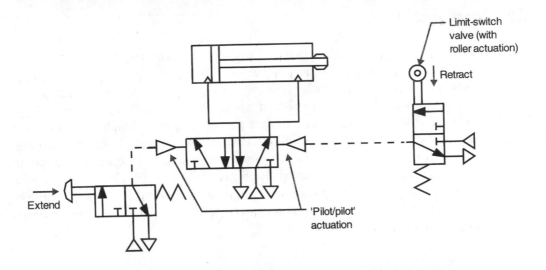

Fig. 10.8 Reciprocating control using a pilot/pilot 5/3 control valve

Typically it is also required to control the speed of an actuator independently in both directions. This is done very simply by using exhaust port flow restrictors (described in Section 9.10) as shown in Fig. 10.9. Speed control obtained by using flow restrictors can be subject to large variations if the applied load or mechanical friction in the actuator or linkages changes for any reason. Designing speed control circuits for consistent and precise control is discussed further in Chapter 15.

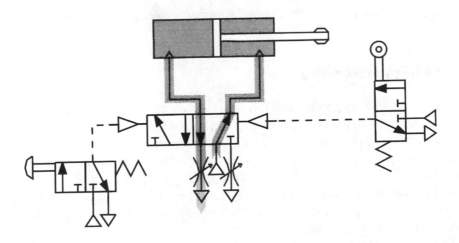

Fig. 10.9 Typical method of controlling actuator speed

10.3 Dwell controls

Most standard pilot operated directional control valves require about 2.5 bar pressure to operate the pilot piston. Depending on the size of the valve, the pilot air quantity may be about 2.0 ml (cm³). If a flow restrictor is inserted in the pilot line, the pilot piston will fill slowly so that the 2.5 bar to switch the valve will be reached after a short time delay. To lengthen the time delay, a small local air reservoir, as shown in Fig. 10.10, can be inserted between the flow restrictor and the valve pilot. Small air reservoirs, in appearance varying from a plugged length of rigid plastic tubing to a cylinder body with no piston or rod, are obtainable as components for use in time delays or as reserve supplies (see Ch. 13). Alternatively, 'timers' or 'time delay valves' are available, incorporating a calibrated flow restrictor and reservoir, into one component.

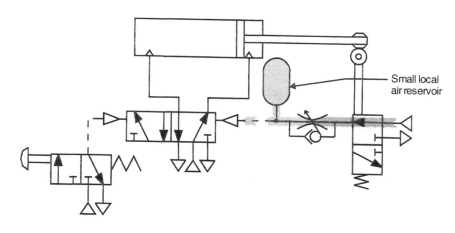

Fig. 10.10 Typical dwell control (delayed return) circuit

10.4 Valve port numbering

When piping a valve into a circuit, it is necessary to use the numbers on the valve body or identification plate as an indication of the function of the ports. The design of valves is so varied that often port function cannot be predicted by the position of a port, particularly where the pilot ports are concerned, which may be adjacent to the work port to be energized by the pilot port, or which may be at the opposite end of the body. Even valves from the same manufacturer may differ in this respect. It is therefore also very common for valve port numbering to be included on circuit diagrams.

Figure 10.11 shows how modern valve ports are numbered. Two-digit pilot port numbers comprise the pressure port (1) plus whichever work port (2 or 4) will be energized or enabled by a pilot signal at that port.

Fig.10.11 Port numbers – present system to ISO5599

Older valves may have lettered ports, as shown in Fig. 10.12.

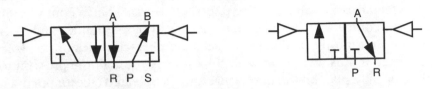

Fig. 10.12 Valve port identification – old alphabetical system

10.5 Identification of components and functions

As mentioned in Section 10.1, components are not always shown on a circuit in their relative working positions, but are arranged so that the circuit is as simple and easy to read as possible. It is also necessary to use the circuit together with function diagrams (see Ch. 12) and component parts lists, so there must be a way of identifying components and actuator movements etc. There is no standardization of functions coding but the system in Fig. 10.13 works well and is used throughout this book. Note that limit-switch valves are shown at a level below the direction

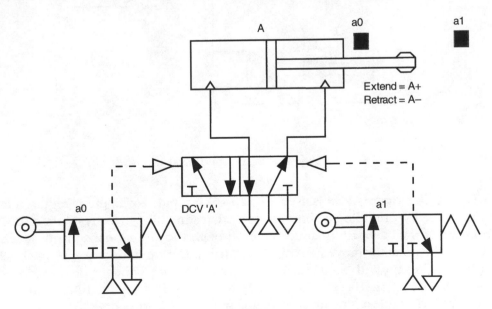

Fig. 10.13 Numbering and identification of components and functions

control valves (DCVs) and are identified by letters and numbers (e.g. a0) so that they can be related to the actual working position adjacent to the actuator. Multicylinder circuits become unintelligible if this way of drawing the circuit is not employed.

10.6 Processing and auxiliary components

The shuttle valve is used where *two* signals or pressure lines are alternatively required to feed *one* actuator or valve port (Fig. 10.14). When a control signal, from a press-button control valve serving a directional valve for example, is not on, it is normally vented to atmosphere due to the standard characteristics of the 3/2 valve. It must be vented in this way to allow air to pass back through the press-button valve when the opposite pilot on the directional valve is operated (see 'trapped' and 'maintained' signals, Ch. 12). If it is required to deliver the control signal from two alternative locations, there must be two press-button valves serving the same pilot port on the directional valve. The shuttle valve is used to prevent the signal from one press-button valve being vented away through the other valve, as shown in Fig. 10.15. It has exactly the same function as the OR valve – see 'logic valves', Chapter 14.

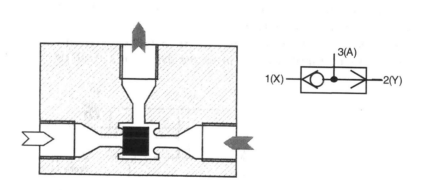

Fig. 10.14 The shuttle valve

The quick exhaust valve is used (Fig. 10.16) if rapid movement is required from an actuator by providing a short-cut for exhausting air, where resistance to flow from the normal air path to atmosphere through the directional control valve will slow-up the actuator movement. A common application of the valve is on winches with a spring-applied band brake, released when the winch motor receives air under pressure (Fig. 10.17). The brake band is pulled 'on' by the return stroke of a spring-return single-acting linear actuator, with work port connected, via the quick exhaust valve, to the winch pressure line. When the motor starts under load, the

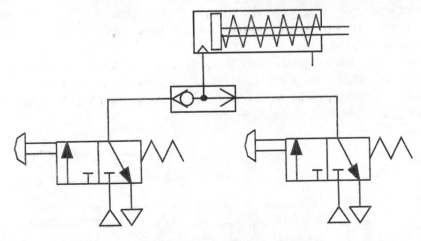

Fig. 10.15 Shuttle valve application

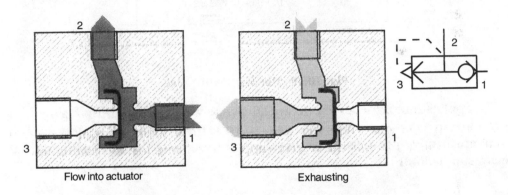

Flow into actuator   Exhausting

Fig. 10.16 The quick exhaust valve

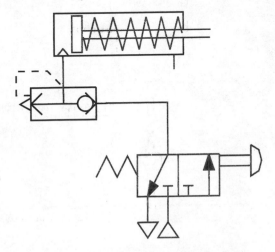

Fig. 10.17 Quick exhaust valve application

cylinder extends, releasing the brake. If load pressure ceases, for example when the load hits the deck or if the air supply should fail, the brake applies automatically and quickly to ensure safety.

Note how the valve symbol gives an explanation of the function: like a shuttle valve with secondary pilot from the work port (2) to block exit via (1).

The **non-return valve** is also commonly called the **check valve** (Fig. 10.18). It allows through-flow in one direction only.

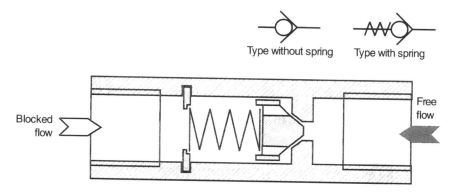

Fig. 10.18 The non-return valve

An application of the valve is to provide a one-way path for air from a main supply into an emergency reservoir (Fig. 10.19). If the main air supply fails, the non-return valve prevents the reserve supply from being lost by venting into the main supply line.

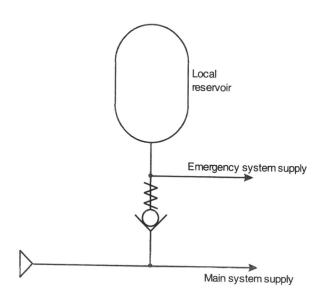

Fig. 10.19 Non-return valve application

The **vacuum generator** or **ejector** uses a supply of air under pressure to create a vacuum source (Figs 10.20 and 10.21). After entering the valve body, the through-flow of air is severely restricted via a nozzle. At the point where air leaves the nozzle, a chamber allows rapid expansion of the air which results in negative pressure locally. The vacuum connection is made into this chamber.

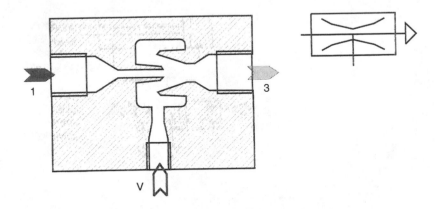

Fig. 10.20 The vacuum generator

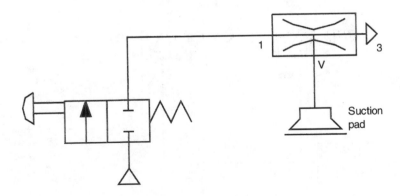

Fig. 10.21 Vacuum generator circuit

The usual application for a vacuum is for the handling or picking up of light-weight components, for example in a packaging machine.

10.7 Circuit layout

It has previously been mentioned in this chapter that:

Components are not always shown in their relative working positions. Positions on the circuit diagram are in the interests of a clear circuit and not to show mechanical relationships.

In application of this, a conventional layout has been evolved, shown here in Fig. 10.22, which assists the production of readable circuits, even when there are many actuators and where logic valves are used to create sequences with repeated or simultaneous functions. Candidates will be expected to use this layout for all the multicylinder circuits they prepare for assessment.

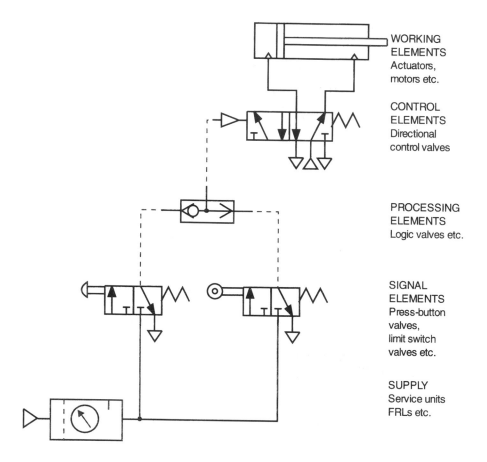

WORKING ELEMENTS
Actuators, motors etc.

CONTROL ELEMENTS
Directional control valves

PROCESSING ELEMENTS
Logic valves etc.

SIGNAL ELEMENTS
Press-button valves, limit switch valves etc.

SUPPLY
Service units FRLs etc.

Fig. 10.22 Circuit layout

Exercises for Chapter 10

1. In conventional pneumatic circuit diagrams:
   (a) What is the difference in representation of pressure and exhaust pipelines in comparison with pilot lines?
   (b) What symbol should be used to indicate joined pipelines?   2.6(a)

2. Draw the diagrammatic circuit using symbols for:
   (a) a single-acting cylinder controlled by one 3/2 press-button valve with spring return;

(b) a double-acting cylinder controlled by two 3/2 press-button spring-return valves;

(c) a double-acting cylinder controlled by a 5/2 pilot/pilot valve which is piloted by two 3/2 press-button spring-return valves;

(d) a double-acting cushioned cylinder controlled by a 5/2 pilot/pilot valve which is piloted by two 3/2 limit-switch valves and started by one 3/2 press-button valve with spring return.   2.5(b)

3. Draw a circuit using a 5/2 double pilot valve, 3/2 push-button valve and 3/2 limit-switch (or 'proof of position') valve to operate a double-acting cylinder with automatic return to the rest position. Which valve would you add to the above circuit to control the outward speed of the cylinder? Draw the symbol and describe the function.   2.5(b)

4. State two functions of reservoirs in pneumatic systems.   2.5(b)

5. The circuit diagram below includes a 'dwell control' comprising a flow restrictor and small reservoir (capacitor), with a sensitive pressure gauge fitted in the pilot connection to the directional valve. Describe the pressure changes you would expect to see on the gauge through the dwell period at the following stages: (a) when the pilot signal first starts; (b) halfway through the dwell delay; (c) when pilot signal results in the directional valve switching.

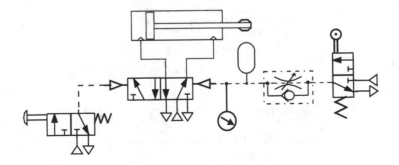

2.5(b)

6. A winch uses a band brake actuated via a linkage by a special single-acting cylinder incorporating a heavy spring to retract the cylinder to engage the brake. Air pressure applied to the full-bore end of the cylinder disengages the brake. Sketch the circuit, showing the cylinder actuated by a 3/2 pilot operated spring-return directional control valve, and include a quick exhaust valve to ensure that when the valve actuation ceases, the brake comes on quickly.

2.6(h)

7. Draw the circuit using symbols for a double-acting cylinder controlled by two pairs of push-button valves, each pair of controls being located at a remote distance from the actuator so that a piloted system must be employed. Draw the circuit with the cylinder in the retracted position.   2.6(b)

8. To the circuit above, add a dwell control so that the action of one remote push-button valve which retracts the cylinder is subject to an adjustable delay.

2.6(b)

9. Draw the diagrammatic ISO1219 symbols for:
   (a) a 3/2 pilot/spring valve
   (b) a lever operated/detent 5/2 directional valve.
   On the symbols indicate:
   (c) the port numbering
   (d) the alternative (obselete) alphabetic port lettering.

2.6(a)

10. Which valve would you add to a circuit to limit the force generated by an air cylinder? Draw the symbol.

2.6(a)

11. Part of a multicylinder circuit is shown below. What is: (a) the name, (b) the purpose of the component off the extension port of cylinder 'A'?
    (c) Why has limit switch valve a* been fitted?
    (d) Describe the cylinder performance during the extension stroke.

2.6(b)

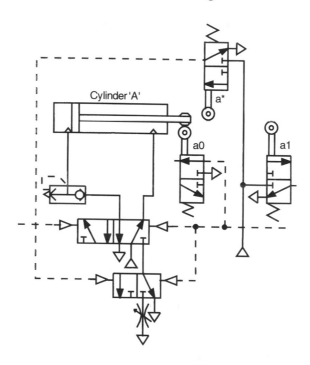

11
Seals and seal systems

11.1 Static and dynamic seals

To seal when there is movement between the surfaces usually requires a different type of seal or seal material, so whether a seal is 'static' or 'dynamic' is key information for the specification of a seal. Figure.11.1 illustrates the difference. Note that the criteria is relative movement of the parts. 'Static' seals can be between two parts which move together. Many 'dynamic' seals are themselves stationary.

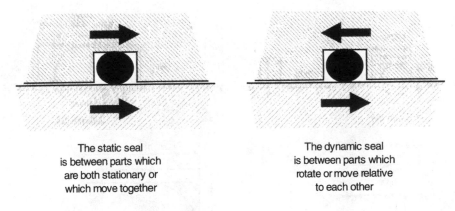

The static seal
is between parts which
are both stationary or
which move together

The dynamic seal
is between parts which
rotate or move relative
to each other

Fig. 11.1 Static and dynamic seals

11.2 Seal types

Static seals are nearly always O-rings: circular section solid nitrile rubber rings. O-rings can also be specified for dynamic sealing at very low speeds, but because they tend to roll and twist this is unusual.

O-rings are produced in standard series of ring diameter and seal surface diameters, but very many standard series have been created in different countries and for varied industrial applications. Since O-rings are not very tolerant of incorrect tolerances in fitting, even for low pressure pneumatic applications, the typical 'O-ring kits' based on one standard series are of very little value for maintenance. The only reliable policy for replacement is to order them specially from the excellent stockists who offer a vast choice, using manufacturers' parts numbers and data or alternatively very careful measurements taken from the part to be replaced.

O-rings may be stretched up to about 20% when fitting and up to 6% to permit them to be used over a larger internal diameter than the nominal ring diameter. In an external groove, a ring may be used which is up to 3% larger than the groove outside diameter. However, rings with diameters larger than an internal groove diameter (except for 'floating' piston seals) or smaller than an external groove diameter should be avoided. O-rings, as with other seals, are available in different 'Shore hardness' ratings. The normal material is 80 Shore A, but a softer more tolerant grade for low pressures is 70 Shore A.

Sometimes for maintenance purposes it is necessary to attempt to determine the size of O-ring required from the groove dimensions, without access to the original seal. Figure 11.2 shows how typical O-rings used as static seals are related to the adjacent sealing surfaces. Use the ring diameter in Fig. 11.2 which is nearest to the actual diameter, or interpolate between the data for the nearest diameters. Note that the groove width allows surplus space for the expansion of the ring when compressed in a sealing situation.

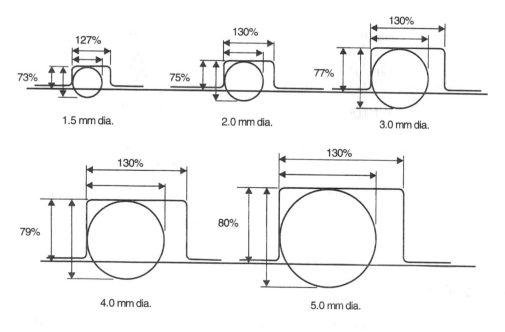

Fig. 11.2 Groove dimensions related to O-ring size for static seals (pneumatic or hydraulic)

Dynamic seals for pistons, cylinder rods or rotating shafts, are usually 'U'-shaped or in a 'Chevron' configuration, so their very low rubbing pressure is increased by the effective fluid pressure when sealing. All seal manufacturers produce different seal types from different mouldings and any standard series of seals is normally unique to that manufacturer, so the correct maker and part number provide the only route to a replacement. This information is often moulded on the seal.

Typical piston and rod seal arrangements can be seen in Chapter 9, Figs 9.1, 9.2 and 9.4. Rod wiper seals are also common on actuators, often in combination with the rod pressure seal. For applications where friction is particularly critical and for actuators in contact with food or where certain aggressive chemicals are present, PTFE seals are common. As PTFE has virtually no elasticity, these seals are usually in the form of a plain ring backed up by a standard rubber O-ring to provide the seal pressure and accommodate working tolerances. Figure 11.3 shows typical seal sections.

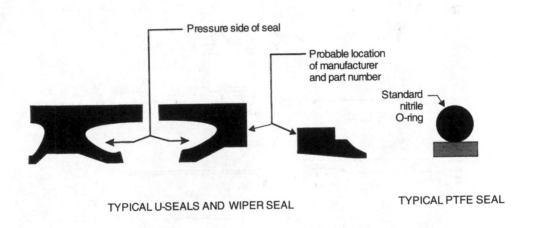

TYPICAL U-SEALS AND WIPER SEAL   TYPICAL PTFE SEAL

Fig. 11.3 Dynamic seal sections

11.3 The installation of seals

It is very easy to damage a seal during installation and a damaged seal will never work for long, if at all. Care in preparation is the key to installing seals.

Seal housings are of two main types, closed and open. To fit a seal into a closed housing it must be snap-fitted, usually by means of a mandrel or fitting sleeve as shown in Fig. 11.5, but sometimes deformed, often by the method shown in Fig. 11.4.

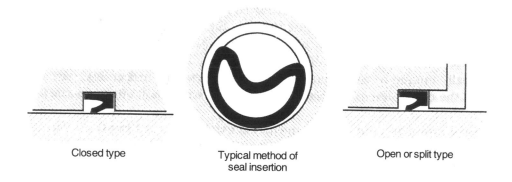

Closed type

Typical method of
seal insertion

Open or split type

Fig. 11.4 Seal housings

Open seal housings can be taken apart to enable the seal to be inserted without the need for deformation.

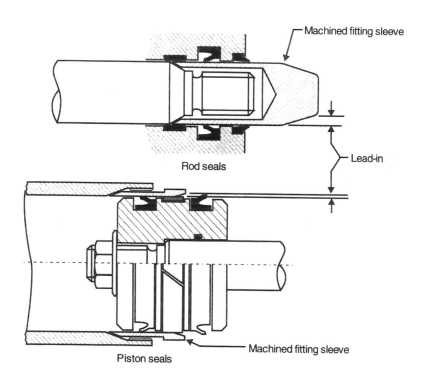

Machined fitting sleeve

Lead-in

Rod seals

Machined fitting sleeve

Piston seals

Fig. 11.5 Fitting cylinder seals

Here are some guidelines for fitting seals:

• Before fitting seals the entire workpiece should be cleaned to remove fabrication residues, filings of metal, and any foreign particles.

- Seals must not be in contact with sharp edges, threads, keyway slots or circlip grooves etc. during fitting.
- Sharp edges should be deburred, chamfered and provided with tapers or radiused corners.
- Never employ sharp-edged tools in the fitting of seals.
- Oil or grease the seal and the housing together with any bores or rod surfaces in the way of the seal fitting.
- Cylinder bores and piston rods must be tapered to avoid damage to the seal.
- For snap fitting of seals, if possible use suitable tools, as shown in Fig. 11.5.

11.4 Seal material

Below is a summary of the most common materials in use.

NITRILE RUBBER

This is the common name for acrylonitrile-butadiene rubber or NBR and is by far the most common seal material – the ubiquitous black rubber. It offers good resistance to swelling in mineral oils, fuels and greases, together with good elasticity, temperature resistance and recovery characteristics after deformation. Nitrile seals can be made with different hardness values, typically measured according to the Shore A scale in the range Shore 60 to Shore 80

POLYURETHANE

This is the common name for polyester-urethane rubber (AU) or polyether-urethane rubber (EU). Polyurethane can be formulated with a greater range of hardness than nitrile rubber and so can be specified where material with a greater hardness than nitrile rubber is required, where a higher mechanical strength is necessary, or where degradation due to the effects of Ozone or Oxygen is important. It is not resistant to some aromatic solvents, brake fluids, acids or alkalis.

PTFE (POLYTETRAFLUOROETHYLENE)

PTFE is a very low friction material with extremely high resistance to almost all chemicals and solvents, even at high temperatures. It has low elasticity and so cannot be used in the same way as rubber or polyurethane for sealing applications, but unlike rubber, the elasticity is little changed at very low temperatures. To seal effectively within normal fitting tolerances where working temperatures are in the normal range, PTFE seals are usually backed up by nitrile rubber rings, a combination which combines elastic resilience with low friction.

OTHER MATERIALS

There are more than 20 organic chemical compounds in regular use for seal materials in addition to the above, developed to provide particular characteristics to suit operating conditions or manufacturing techniques. Each of these may have as many as eight alternative brand names from different manufacturers. Great care is necessary when specifying a seal material to ensure that the operating conditions and material properties are thoroughly known and that they are fully compatible.

A common proprietary type of fluorine rubber seal material specified for resistance to high temperatures is called 'Viton'. If found in a charred or sticky condition as a result of very high temperatures, this material is dangerous and should be handled using special procedures as advised by the equipment or seal manufacturer.

Exercises for Chapter 11

1. Briefly describe the difference between static and dynamic seals.   2.5(f)

2. Sketch an external O-ring groove in section, with the ring shown in the groove without the adjacent part, so that the seal is not deformed into the groove, indicating the approximate relative dimensions of the groove and ring.   2.5(g)

3. Sketch:
   (a) a typical single-acting U-shaped dynamic seal in section;
   (b) a typical two-part PTFE seal assembly in section;
   (c) double-acting piston seals in section, showing the two single-acting U-shaped seal elements and the wear ring.   2.5(g)

4. Explain briefly with the help of a sketch the difference between a closed seal housing and an open seal housing.   2.5(g)

5. List five of the precautions which should be adopted when fitting seals. 2.5(f)

6. Describe briefly three common standard construction materials for seals, giving an outline of the material properties and stating typical applications.   2.5(f)

7. State two properties in which the behaviour of PTFE differs from other common seal materials.   2.5(f)

8. Give one way (not including the machine manufacturer's documentation or spares provided) of determining the specification of:
   (a) an O-ring seal
   (b) a U-shaped piston seal.   2.5(g)

9. Outline (a) the purpose, (b) the position in a cylinder, and (c) the type (i.e. whether static or dynamic) for each of the following seals:
   (i)   rod wiper
   (ii) O-ring body seal
   (iii) bidirectional piston seal.   2.5(f)

12
Pneumatic circuits (2)

12.1 Simple multicylinder circuits

Chapter 10 concluded in Section 10.7 with recommendations for the layout of circuits. In this chapter and Chapter 16, the advantages of those methods will be clear, otherwise multicylinder circuits are much more difficult to design and to understand.

Some basic features of many circuits and of the circuits presented here are the **sequence** and the **function diagram**. Figure 10.13 introduced ways of identifying actuators and actuator functions. The sequence, e.g. A+ B+ A− B−, is a way of writing the machine functions using this method of coding – see Figs 12.1 and 12.2.

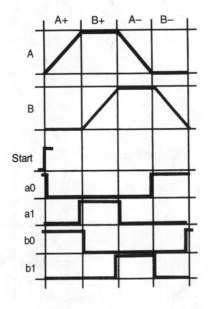

Fig. 12.1 The function diagram

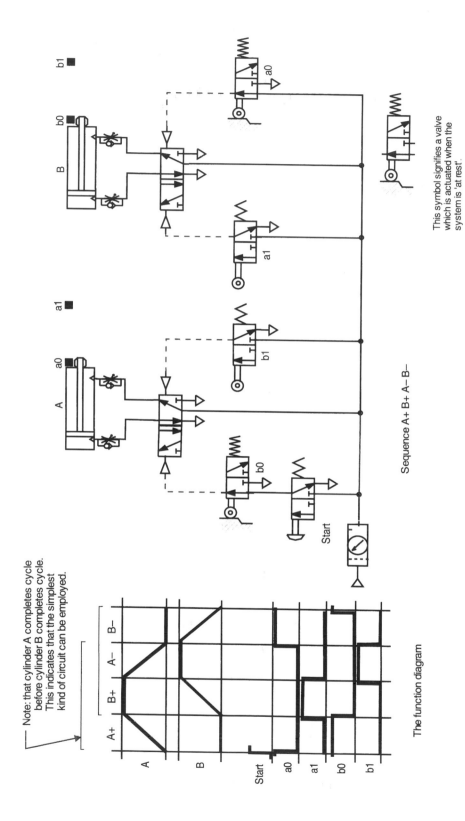

Fig. 12.2 A simple two-cylinder circuit

Note: that cylinder A completes cycle before cylinder B completes cycle. This indicates that the simplest kind of circuit can be employed.

This symbol signifies a valve which is actuated when the system is 'at rest'.

Sequence A+ B+ A− B−

The function diagram

The function diagram is a step-by-step graph illustrating the sequence for each actuator, and the essential first step in designing a circuit. Optionally the functions of some or all of the valves can be added beneath the actuators. When the circuit is working correctly the diagram has very little use – it gives information which is better shown by the machine in operation. If the machine breaks down then the function diagram is extremely useful. Machines seldom break down and stop in the 'at rest' position. Interestingly, even machine operators or maintenance staff often cannot explain exactly what a machine does when it has stopped. Even manufacturers sometimes overlook this basic information in their manuals. Since the function diagram shows the position of each working part at any point, it can be used in a breakdown situation to indicate the stage in the sequence at which the machine stopped, together with the preceding and following functions, which assists towards a rapid analysis of the cause of the problem (see Ch. 19).

The valves which are actuated when the circuit is at rest are shown by the modified valve symbol in Fig. 12.3. Note that the valve is drawn with the connections to the position which is actuated when the circuit is at rest, not the usual convention (as explained in Ch. 9) with connections to the position when the valve is at rest.

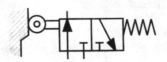

Fig. 12.3 Symbol for valve actuated 'at rest'

The two-cylinder circuit of Fig. 12.2 is simple because the sequence of operations does not generate any conflicts between opposing pilot signals across the directional control valves. Follow the functioning of the circuit (with reference to Fig.12.2) through each step as follows:

- The start signal goes through valve b0 (actuated by B at rest) to pilot the directional control valve (DCV) to extend cylinder A. Cylinder A can extend because valve b1 is not actuated.
- When cylinder A hits valve a1, B is free to extend in the same way because valve a0 is not actuated.
- When cylinder B hits valve b1, A is free to retract because valve b0 is not actuated (since B is still extended).
- When cylinder A hits valve a0, B is free to retract because A is no longer in contact with valve a1.
- When B retracts, valve b0 is actuated providing an air path for a start signal to extend A.

Put another way (with reference to the function diagram), the function of neither actuator is enclosed within the function of the other actuator, telling us that the simplest circuit will deliver the sequence required.

In Fig. 12.4, on the other hand, the function diagram illustrates a sequence in which the function of cylinder B is enclosed within the function of cylinder A. It has been drawn as a simple circuit, but it will not work, due to the occurrence of **trapped signals** and **maintained signals**. For example:

- The start signal goes through valve a0 (actuated by A at rest) to pilot the directional control valve (DCV) to extend cylinder A. Cylinder A cannot extend because valve b0 is actuated, resulting in an opposing or maintained signal which is applying pilot pressure to the other end of the DCV. The result is that nothing will happen.
- The situation would be repeated when B tried to retract.

Definitions

*A pilot signal which has completed its function and then remains effective so that it blocks the effect of other signals is called a **maintained signal***

*A pilot signal applied to a valve with an opposing pilot signal also effective is called a **trapped signal***

12.2 Pulsed signals

If the maintained signals from limit-switch valves in Fig.12.4 were short or 'pulsed' signals, no trapped signals would occur and the circuit would work. This can be achieved with the 'one-way trip' (or 'idler roller') valve shown in Fig.12.5 and the circuit in Fig.12.6, or by using pulse generators as in the circuit in Fig.12.7.

Definition

*A pilot signal which is made to occur for a short duration only and which then reverts to an exhaust path is called a **pulsed signal***

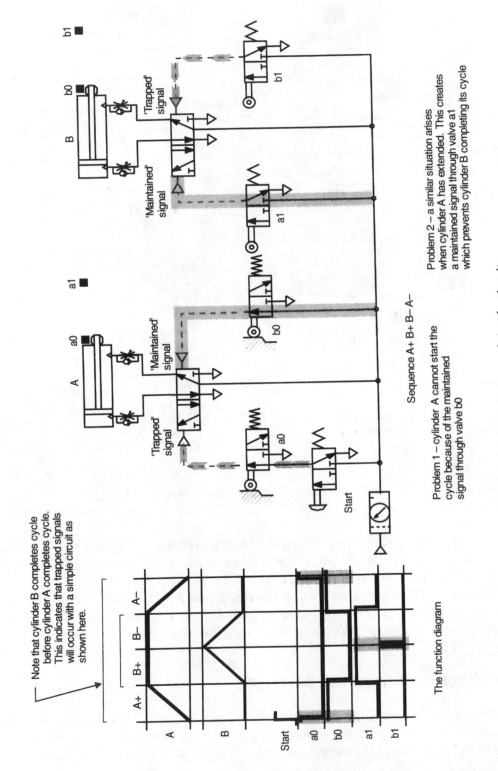

Note that cylinder B completes cycle before cylinder A completes cycle. This indicates that trapped signals will occur with a simple circuit as shown here.

'Trapped' signal

'Maintained' signal

A   a0

'Maintained' signal

'Trapped' signal

B   b0

a1

b1

Start

a0

b0

a1

b1

Sequence A+ B+ B– A–

Problem 1 – cylinder A cannot start the cycle because of the maintained signal through valve b0

Problem 2 – a similar situation arises when cylinder A has extended. This creates a maintained signal through valve a1 which prevents cylinder B completing its cycle

A+   B+   B–   A–

A

B

Start

a0

b0

a1

b1

The function diagram

Fig. 12.4 The limitations of simple circuits

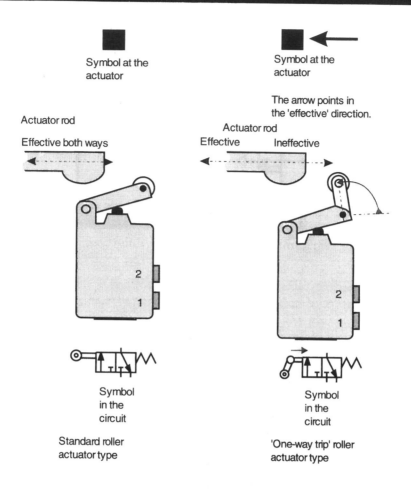

Fig. 12.5 Pneumatic limit switches

Note that idler roller type limit switch valves generate a pulsed signal when actuated from one direction and no signal when actuated in the opposite direction, as indicated by the arrow with the symbol at the working location.

Idler roller type limit switches have a reputation for unreliability. Pulse generator valves are a 3/2 pilot/spring valve piloted off via a delay restrictor, or the equivalent in one component: a more robust way of generating pulses.

12.3 Circuits with repetitive movements

Repeated movements of an actuator are difficult to provide with standard pneumatic circuits unless pulse generators are employed in conjunction with a **counter**, allowing the pilot signal through after a set number of pulses (see Fig. 12.8).

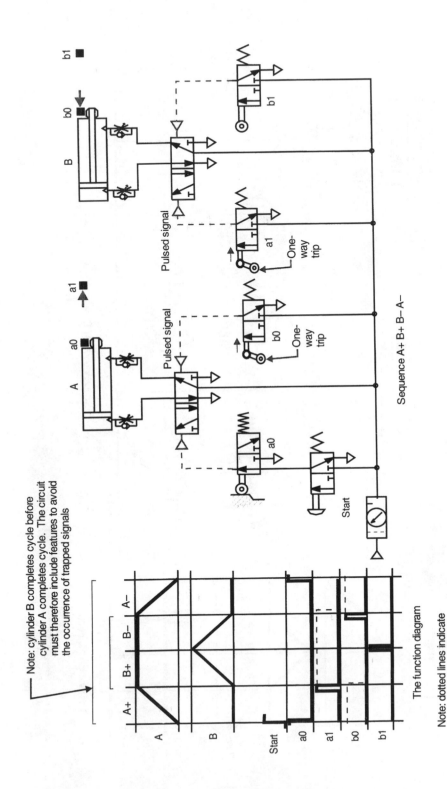

Note: cylinder B completes cycle before cylinder A completes cycle. The circuit must therefore include features to avoid the occurrence of trapped signals

Pulsed signal

Pulsed signal

One-way trip

One-way trip

Start

Sequence A+ B+ B– A–

The function diagram

Note: dotted lines indicate contact with the limit-switch. Firm lines show resulting signal

Fig. 12.6 One way to avoid trapped signals

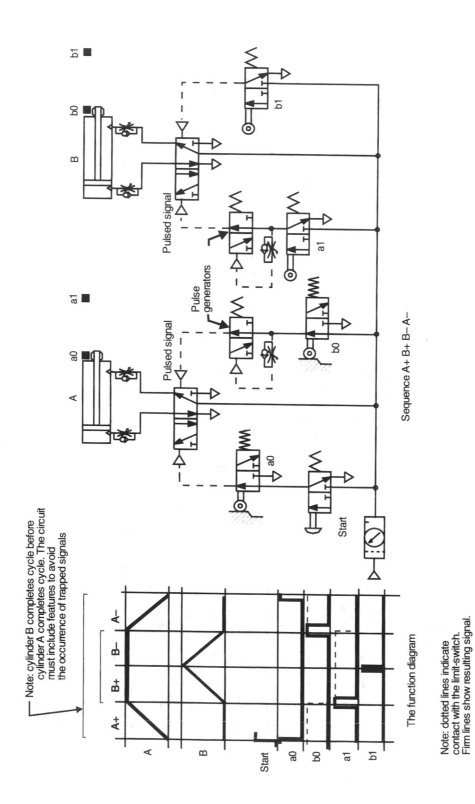

Note: cylinder B completes cycle before cylinder A completes cycle. The circuit must include features to avoid the occurrence of trapped signals

The function diagram

Note: dotted lines indicate contact with the limit-switch. Firm lines show resulting signal.

Sequence A+ B+ B– A–

Fig. 12.7 A reliable way to avoid trapped signals

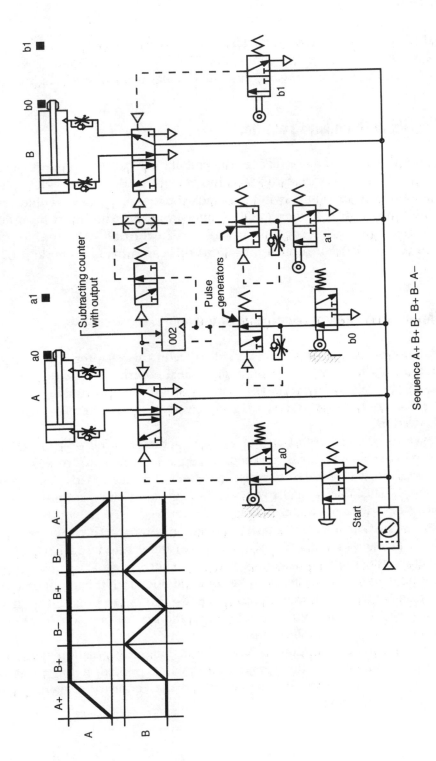

Fig. 12.8 A pulsed signal and counter circuit for repetitive movements

12.4 Latching circuits

To avoid trapped signals, with the addition of another directional control valve, a circuit can 'latch' between two zones (Fig. 12.9). This reliable system with a low component count is a step towards cascade circuits, described in Chapter 16.

12.5 Circuits with no limit switches

When a lightly loaded actuator reaches the end of its stroke, the inlet pressure rises steeply. The rise in pressure can be used to actuate a valve to provide a pilot signal in the same way as a limit switch. A standard pilot/spring valve would respond at too low a pressure, so either an adjustable spring pressure must be used, or a valve with unequal pilot sections called a **differential pilot valve** can be employed (Fig. 12.10). For some applications this can provide a simple robust circuit.

12.6 Three-position directional control valves

In typical circuits like all those described so far in this chapter, each actuator is associated with a two-position piloted directional control valve. After receiving a pilot signal, the valve remains in the same position whether the signal is maintained or not, until an opposite signal is received to switch the valve back to the original position.

However, for the control of a double-acting actuator, three position DCVs are common where the actuator is required to stop at intermediate strokes, or where for safety reasons it is required to resist the effect of applied loads if, for example, the air supply fails. The typical five-port, three-position valve is known as a 5/3 valve and is described in greater detail in Chapter 14.

The three-position valve is actuated to positions one and three (corresponding to positions one and two of a two-position valve) and is fitted with springs at each end to centre the valve at position two when there is no actuation. Typically, valves are pilot actuated but may also be solenoid operated or lever operated with either 'spring centring' or 'three-position detent'. Valves can be fitted with spools (see Ch. 14) in which all ports are closed in the centre position or alternative arrangements for special applications.

Figure 12.11 shows a circuit for actuating a hinged ventilation window. The valve supplies pressure to the actuator only when piloted. If the press-button spring-return pilot valves are not actuated then the ventilator remains parked at any intermediate position.

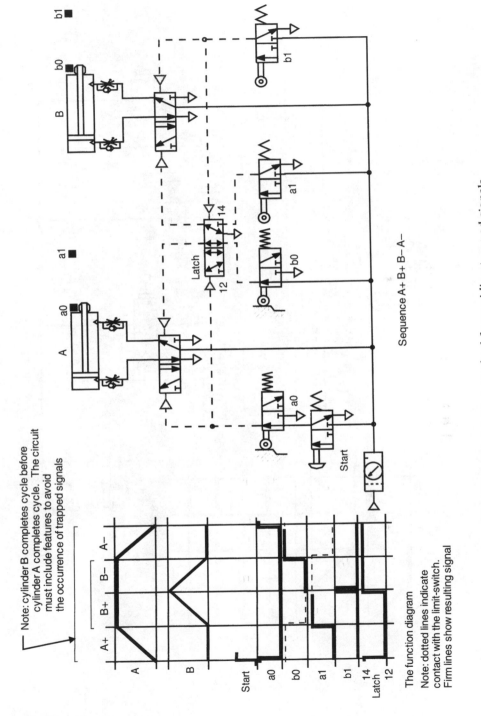

Note: cylinder B completes cycle before
cylinder A completes cycle. The circuit
must include features to avoid
the occurrence of trapped signals

Sequence A+ B+ B– A–

The function diagram

Note: dotted lines indicate
contact with the limit-switch.
Firm lines show resulting signal

Fig. 12.9 The latching method for avoiding trapped signals

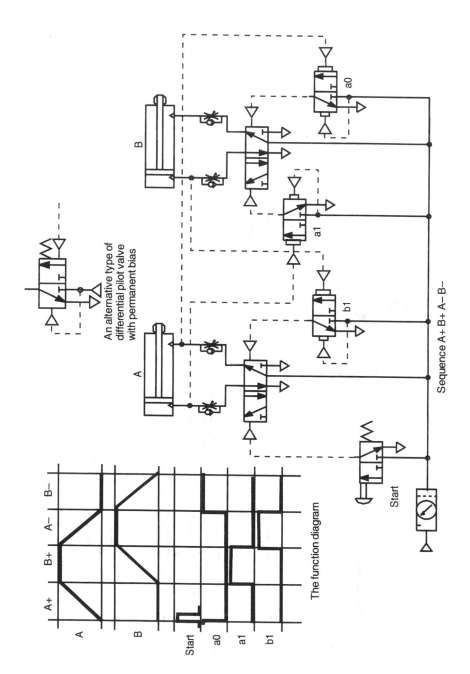

Fig. 12.10 Control with pressure signals and differential pilot valves

An alternative type of
differential pilot valve
with permanent bias

The function diagram

Sequence A+ B+ A− B−

Start

A+   B+   A−   B−

A

B

Start

a0

a1

b1

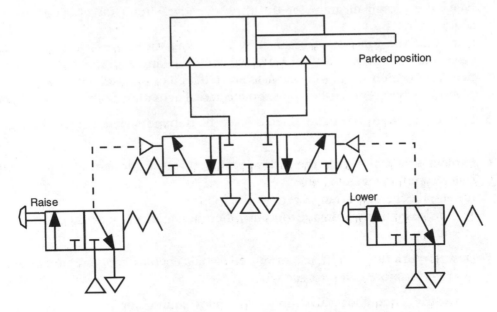

Fig. 12.11 Use of 5/3 valve as a ventilator control

Exercises for Chapter 12

1. Draw the conventional ISO1219 symbol for a 3/2 roller operated/spring-return limit-switch valve, indicating the connection points:
   (a) for a position in the circuit where it will be actuated during the machine sequence;
   (b) for a position in the circuit where it is actuated before the sequence starts (i.e. when the machine is 'at rest'). 2.6(a)

2. Consider a three-cylinder circuit which:
   (i) raises a heap of waste material to a work area and holds it there;
   (ii) compresses it sideways then withdraws;
   (iii) compresses it lengthwise then withdraws;
   (iv) lowers it back to the position at which it started.
   State the cylinder movements by
   (a) a function (step motion) diagram for actuators only;
   (b) notation (i.e. the sequence as A+ B+ etc.). 2.6(f)

3. (a) Draw a function diagram for the actuator movements only of a three-cylinder circuit with the sequence A+ B+ C+ A– B– C–.
   (b) Will this sequence require a special circuit to be devised to avoid maintained or trapped signals? 2.6(f)

4. Draw the circuit diagram for the three-cylinder circuit (as above) with the sequence A+ B+ C+ A− B− C−.
Complete the function diagram now by adding the movements of the start valve and limit switches a0 a1 b0 b1 c0 and c1, taking care to position the vertical lines which show where valves are actuated or released close to the ends of stroke, representing the approximate actual actuation points. 2.6(f)

5. Define the terms: (a) maintained signal, (b) trapped signal, (c) pulsed signal. 2.6(f)

6. Explain in a few words the following functions of a 'one-way trip' or 'idler roller' type limit switch valve:
(a) the mechanical arrangement of the roller;
(b) the resulting pneumatic output when actuated in one direction or the other. 2.6(f)

7. Describe briefly why the use of pulsed signals enables very simple circuits to be used for more complex sequences. 2.6(f)

8. Draw the circuit for valve only with pilot source for a 'pulse generator'. Describe briefly how the assembly works and how pulse duration is adjusted. 2.6(f)

9. The part function diagram alongside shows the limit-switch valve movements only of a 'latching circuit'. What is signified by the use of alternatively solid or dotted lines? 2.6(f)

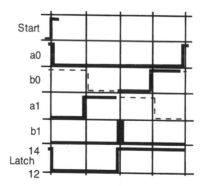

13
Emergency and fail-safe systems

13.1 Safety

Air is a safe fluid but can be dangerous when compressed, due to the stored energy. It expands explosively if suddenly released or if the container fails. Compressed air remaining in receivers, pipelines or components after shutdown can therefore be a major safety hazard. Most machines can be dangerous, but an apparently inert machine which contains stored energy is doubly dangerous.

The pressure systems regulations (see Ch. 8) cover the ratings, operation and maintenance of all but very small systems, but remember that the pipework and minor components of systems, not in themselves subject to the regulations, may also constitute a safety hazard. Small systems outside the scope of the regulations can also cause injuries. Additional important factors to be considered in the promotion of safety are:

- The provision of emergency shutdown devices and procedures which stop the machine and if necessary ensure that relevant parts of the system are vented. These are called **emergency stop circuits**.
- Systems which are 'fail-safe', so that failures or misuse do not endanger personnel or equipment. These are called **fail-safe circuits**.
- Operating procedures which discourage misuse of equipment. A common example is **two-handed starts**.

The following selection of simple circuits are representative solutions but should be checked for suitability before applying to any particular system.

13.2 Emergency stop circuits

Figure 13.1 is a circuit for control of a conveyor. Within the limited capabilities of an air operated system, it ensures that the conveyor (which may be carrying heavy goods travelling at speed) stops immediately and remains stopped if loads are then moved. An emergency reverse is also provided. Note that the two press-button 3/2 valves differ from each other: the 'reverse' valve is the usual 'normally closed' type, but the emergency stop valve is a 'normally open' type which blocks the passage of pressure when actuated.

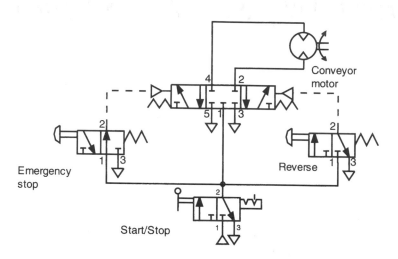

Fig. 13.1 Emergency stop circuit for a conveyor

A machine that raises a load, for example, can be designed to lower the load slowly if the emergency stop is pressed. Figure 13.2 shows a similar situation which arises with a baling press. When the press is stopped normally, the force

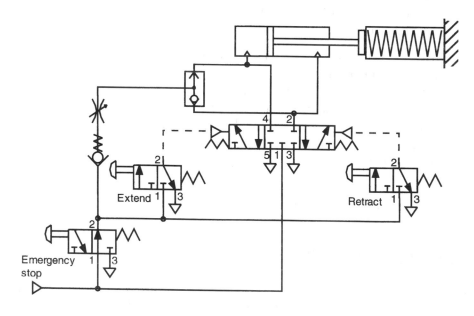

Fig. 13.2 An emergency stop and vent circuit for a baling press

from the part-compressed bale creates locked-in pressure, but if the emergency stop is pressed, the cylinder is vented via a restrictor and non-return valve to enable the cylinder to move and dissipate the pressure.

13.3 Fail-safe circuits

A machine may stop working because the main air supply has failed at a position in the sequence where the load or workpiece could constitute a danger to personnel. The design should provide for this with an emergency supply together with automatic functions to return the load to a safe state.

Figure 13.3 shows a circuit incorporating a small emergency reservoir which is arranged to return the cylinder to the retracted position. The directional control of the actuator is by means of two 3/2 pilot/pilot valves. When the main supply pressure drops below a level set by the adjustable pilot/spring 3/2 valve, the local reservoir takes over to pilot the directional valves to use the emergency supply. When the mains supply pressure is restored, the directional valves remain in the emergency position until reset. Note that the directional valves are drawn with double arrows for the air path, showing that the 'exhaust' ports (3) are in this instance emergency inlet ports. Most, but not all, pneumatic spool valves can accept air in the 'wrong direction'.

Figure 13.4 shows the same system with 'automatic reset'. The unbalanced pilot areas on the 'differential pilot' (see Ch. 12) directional valves give automatic priority to the reset signal.

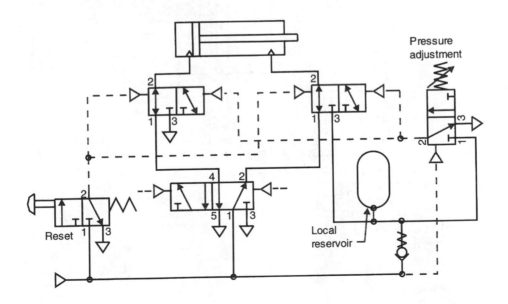

Fig. 13.3 A fail-safe circuit. Mains failure results in the actuator retracting

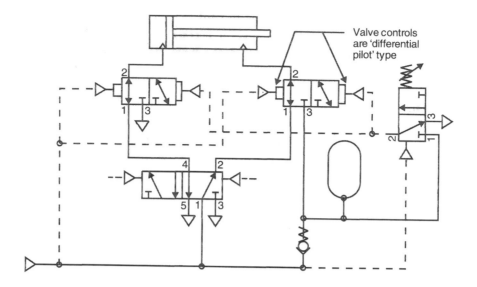

Fig. 13.4 A fail-safe circuit similar to Fig.13.3 but with automatic reset

Figure 13.5 shows a drill circuit in which the workpiece clamp and guard are operated by a foot pedal control. If the guard is detected in the safe working posi-

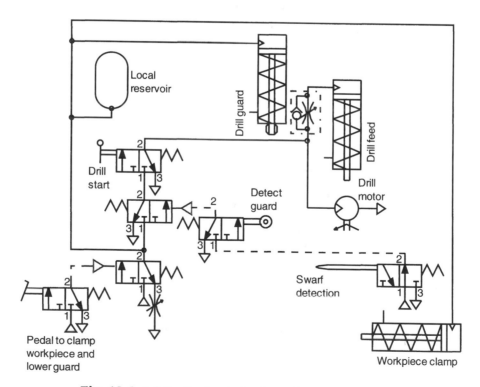

Fig. 13.5 A fail-safe circuit for the safe control of a drill

tion then the supply is switched to the lever/spring start valve for the drill motor and drill feed. If the guard should move from the safe position, or swarf is detected on the drill bed, then the supply is removed. However, a local reservoir ensures that in this case, or in the case of a main supply failure, the drill motor stops before the clamp pressure is released or the guard retracts. This is a typical fail-safe system, providing for both malfunctions and misuse.

13.4 Two-handed controls

Precautions may be necessary to ensure that an operator does not use potentially hazardous short-cuts such as pulling down a drill feed with one hand while starting the drill with the other, or opening a storeroom door while steering a pallet truck. This can be arranged with two controls requiring the use of both hands. Widely separated two-handed controls are a way, for example, of ensuring that a cold room door cannot be opened by an unaccompanied person. Figure 13.6 shows common circuits using 3/2 'normally open' pilot valves. The 'AND gate' used in the parallel circuit is described in Chapter 14.

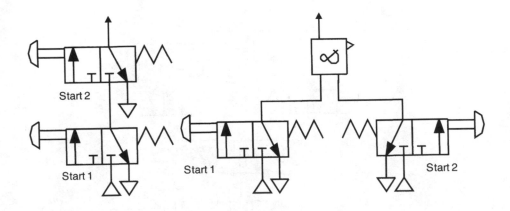

Fig. 13.6 Circuits for two-handed controls

13.5 Blocking or lock-up valves

This is a useful safety device for mounting in a delivery or actuator supply line, which closes immediately the system pressure drops below a predetermined value and re-opens only if the set pressure is re-established. A sudden change in the position of cylinders under load due to a sudden drop in system pressure can be prevented, which offers the possibility to maintain a process control system with the valves fixed in the momentary position until the pressure is recovered. Figure 13.7 shows an application of the dual version.

Available forms of this valve are:

• two-port two-position on/off configuration, similar to a pilot/spring 2/2 valve with adjustable spring pressure from 1.4 to 7 bar pilot signal;
• a double version of the 2/2 type above suitable for closing a pair of pipes simultaneously;
• three-port two-position configuration, similar to a pilot/spring 3/2 valve with adjustable spring, to allow switching to a secondary supply source when the signal pressure drops below preset value.

13.6 Unloading or residual pressure release valves

This valve is an emergency device to enable maintenance personnel to vent residual air pressure or stored energy from a locked line. Basically similar to a 3/2 press-button spring-return valve, it is designed for direct push-in installation in local pipelines, and without any maintenance will remain ready for use near the actuator at any time when venting may be required for safety reasons. Figure 13.7 shows two valves in actuator supply lines which are blocked when the DCV is not piloted or due to closure of the blocking valve (see Section 13.5 above).

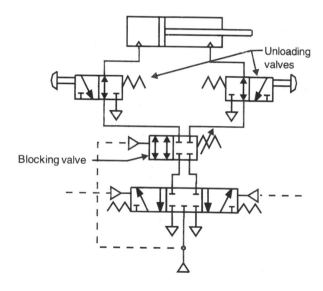

Fig. 13.7 Application of blocking and unloading valves

Exercises for Chapter 13

1. Outline the meaning of the terms: (a) fail-safe system, (b) emergency stop system, (c) Two-handed control.
   2.7(a)

2. Briefly explain the action of a blocking or lock-up valve.   2.5(b)

3. Briefly explain the function of an unloading or residual pressure release valve.
                                                                      2.5(b)

4. Describe the sources of pressure that may remain in a pneumatic system after
   the air supply has been shut down and vented.   2.9(c)

5. State which of the circuits (a) and (b) in the figure below is the 'fail-safe system' and which is the 'emergency-stop system'. For each system outline the
   purpose and function of the valve 'X'.   2.7(b)

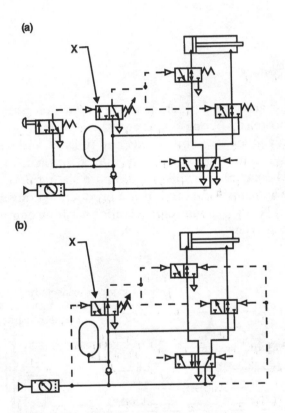

6. A moulding machine in which plastic bottles are made and then air-tested is
   manned by two operators who are paid according to the number of bottles
   produced. One man could operate the complete process, but for safety reasons
   this must never occur. Which of the following safety devices is likely to be
   effective:
   (a) a fail-safe circuit with automatic reset from a local reservoir;
   (b) automatic shutdown unless widely separated buttons are pressed together
       when a bleeper sounds every 3 minutes;
   (c) a double blocking valve to safeguard continuity of machine functions when
       supply fluctuates due to concurrent demands of other machines.   2.7(b)

14
Pneumatic components (2)

14.1 Rodless cylinders

The standard linear actuator with an actuator rod which projects from the cylinder is a simple and reliable component and the best form for most applications. However, rodless cylinders are now sturdy and reliable, and sometimes, if there is not enough room in a machine for the cylinder to project to one side, a rodless type must be used. Examples of this situation are sliding door actuators in a confined space or the actuators mounted over a production conveyor for the purpose of spraying paints, placing portions of food or for washing components. Figure 14.1 shows two common types.

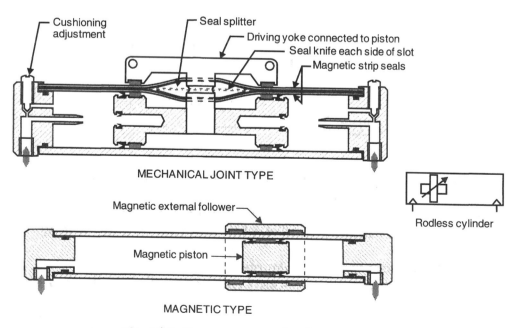

Fig. 14.1 Common types of rodless cylinders

14.2 Cylinders with limit switches

Electrical limit switches are typically fitted to the outside of a cylinder and operated magnetically from inside moving parts. Known generally as **proximity
switches**, they may be in the form of adjustable clamps added to a standard body
or integrated into the cylinder design as manufactured. The simplest form of switch
is the **reed switch**, a magnetic leaf switch enclosed in a small glass inert gas-filled
phial, actuated by a piston magnet. **Solid state** inductive switches are similar in
appearance but offer faster operation and greater accuracy. Figure 14.2 shows a
typical reed switch arrangement.

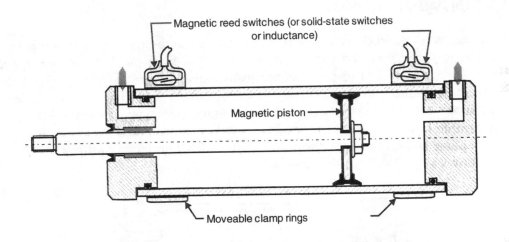

Fig. 14.2 Double-acting cylinder with magnetic proximity switches

14.3 Non-rotating cylinders

The rod of a standard linear actuator can be rotated. This causes the piston to
rotate, with resulting friction which leads to seal and bearing wear and also to a
risk of the rod becoming detached from the piston. Rotation of the rod of a standard actuator must therefore be prevented by allowing for rotational movement
within the mechanism of the machine and at the same time restraining the rod so
that it cannot rotate. In addition, side loads should not be imposed on the rod of a
standard linear actuator unless additional external bearings are provided. Non-
rotating cylinders incorporate restraint against rotation together with additional
bushes capable of withstanding some side loads, so they provide an economical
package for the machine designer where these characteristics are required. Figure
14.3 shows a typical non-rotating cylinder.

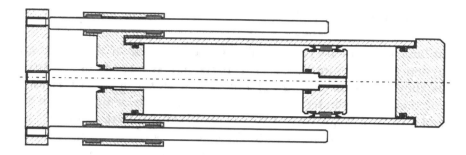

Fig. 14.3 Cylinder with 'H' guides

14.4 Through-rod cylinders

Cylinders with the rod projecting from each end may be required for two principal reasons: to enable the rod to be linked with a hydraulic cylinder for speed control (see Ch. 15), or to equalize the piston areas on each side to prevent the actuator creeping. Creeping occurs due to the unbalanced areas on each side of the piston when both cylinder ports are connected and some pressure is effective, for example if the exhaust ports are in an exhaust return circuit as found in some recent plant installations for the purpose of improving energy efficiency. Figure 14.4 shows a typical through-rod cylinder.

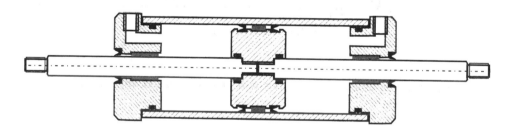

Fig. 14.4 Through-rod cylinder

14.5 Diaphragm cylinders

Very short stroke cylinders are often of similar construction to standard cylinders, but they can also be constructed with a rubber or metal diaphragm in place of the conventional piston. Diaphragm cylinders have the advantage of zero leakage between the inlet and discharge ports and much lower levels of friction than standard cylinders. Another form of single-acting short-stroke actuator comprises one or more rubber doughnuts stacked together which provide longitudinal force and displacement as they are inflated. Figure 14.5 shows a typical single-acting spring-return diaphragm cylinder.

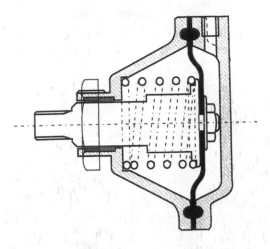

Fig. 14.5 Single-acting diaphragm cylinder

14.6 Rotary actuators

Where a twisting or turning function is required, the rotary actuator, a motor with part-turn rotation only, is often employed. Suitable for actuating process control valves where the maximum force required is usually at the start of the stroke, they are designed to give very high starting torque.

Figure 14.6 shows the **rack and pinion actuator,** available with strokes up to one turn if required, which can include cushioning as illustrated. Some recent types have four pistons with integral racks grouped in a square formation around a central shaft, which give very high torque from a small unit.

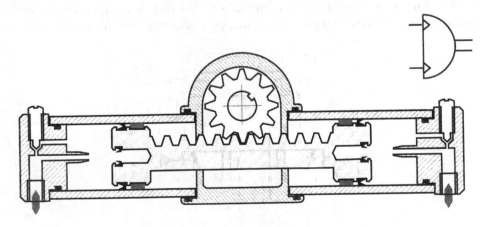

Fig. 14.6 The rack and pinion rotary actuator

For strokes of a quarter-turn or less, the **vane actuator** offers good torque capability and constant torque through the full stroke. Versions are also available with built-in return springs, as required in fail-safe systems for example, three positions or proportional position control and integral linkages to give 180° rotation. Figure 14.7 illustrates the basic design, with only one moving part.

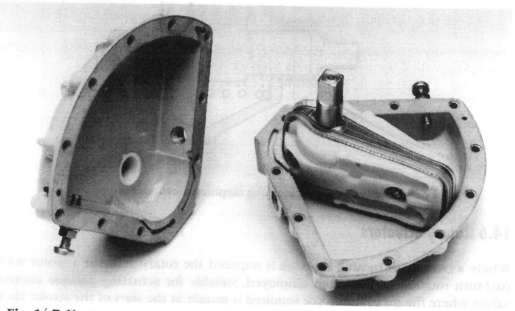

Fig. 14.7 Vane type rotary actuator. (Photograph by kind permission of Kinetrol Ltd.)

14.7 Cylinder locking devices

To rigidly lock an air cylinder, either hydraulics must be used, as explained in Chapter 15, or a mechanical brake, spring applied and released by air pressure, can be added to the cylinder. Figure 14.8 shows a common circuit for non-electrical

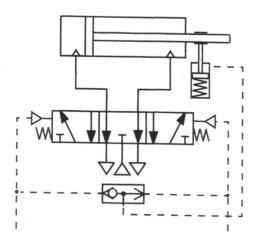

Fig. 14.8 Locking device symbol and typical circuit

systems. Note that in the centre position, the 5/3 valve vents both sides of the cylinder, otherwise minor leakages could result in pressure build-up in the cylinder, causing violent movement at the moment of release. Figure 14.9 shows a typical cylinder locking device.

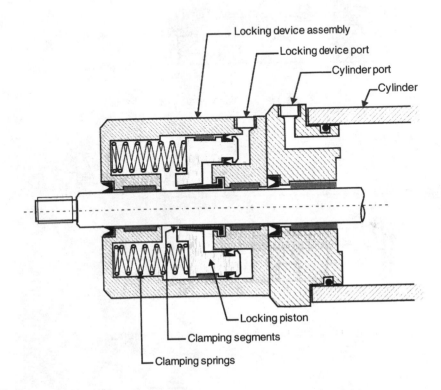

Fig.14.9 Typical cylinder locking device

14.8 5/3 directional control valves

Chapter 12 introduced the three-position valve actuated either side of spring-centre position. Five-port valves with alternative spool types to suit application requirements are shown below. Valves are shown as air–pilot operated but may be solenoid or lever operated. Two-digit pilot port numbers comprise the pressure port (1) plus whichever work port (2 or 4) will be energized.

Figure 14.10 shows the type with all ports closed in centre position. When not actuated, this valve traps the air at both sides of the actuator piston or motor. Figure 14.11 shows the type with work ports vented in centre position. Figure 14.12 applies pressure to both sides of the piston, so a standard single-rod cylinder will extend with a small force resulting from rod area only, a useful feature where a small clamping or 'nip' force is required.

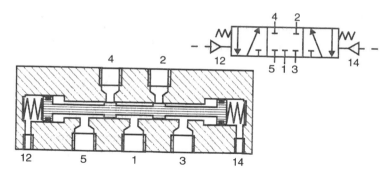

Fig. 14.10 5/3 Valve – all ports closed

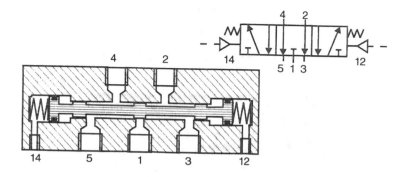

Fig. 14.11 5/3 Valve – work ports vented

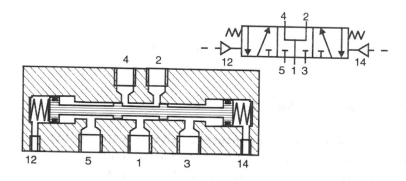

Fig. 14.12 5/3 Valve – work ports pressurized

14.9 The programme sequencer

As explained in Chapter 12, some multicylinder sequences require circuits using pulsed signals or latching between two or more zones to avoid the occurrence of 'maintained' and 'trapped' signals. The latching circuit could be said to 'step' from zone to zone, and the pilot/pilot two-position valve which latches the supply and remains in position until it receives a subsequent signal could be said to have a 'memory' of each mode in which the circuit works.

Chapter 16 describes various design solutions for more complex multicylinder circuits, including 'stepper' circuits in which each step in the sequence forms a zone in which the function must be completed before stepping to the next zone. Built with conventional pneumatic components, 'stepper' circuits can be complex and expensive, but if components specially developed for this type of circuit are used, i.e. sequencers, the result is lower costs, simple layout, simple design and less difficulty when troubleshooting.

The basic component of all-pneumatic stepper circuits is the **programme sequencer**, a stackable module including a special type of pilot/pilot 3/2 valve (the memory) together with an AND logic valve (see Section 14.10) and an OR valve (shuttle valve), which when assembled in a chain of modules steps through the circuit functions, enabling each function to be handled independently.

Programme sequencers are interconnected via a multistage subplate and each module has a mounting face for additional logic valves if required.

Figure 14.13 is an approximate diagrammatic representation of the circuit comprised in two sequencer modules. Each module functions as follows:

- At rest – the spool is biased to position A. Main pressure is blocked.
- Start – the spool is piloted to position B. Pressure goes to the action port, reset signal goes to previous module via the shuttle and is applied to one input port of the AND valve.
- Action – spool remains in position B (start signal may be removed).
- Verification – action completed signal to other input port of AND valve which sends start signal to next module. Reset signal arrives from next module to pilot spool to position A. The action port connects to exhaust. Spool now at rest.

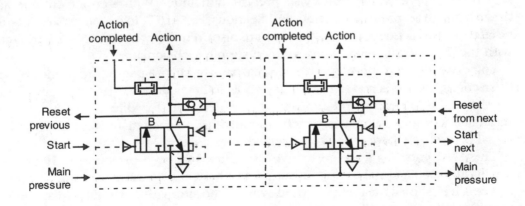

Fig. 14.13 Circuit for two programme sequencers

Figure 14.14 is a slightly simplified section through a typical programme sequencer module showing the mechanisms for carrying out the stepper functions described. The functioning of stacked programme sequencers through a sequence is covered in Chapter 16 in Figs 16.8–16.10 and in typical circuits in Figs 16.12 and 16.13.

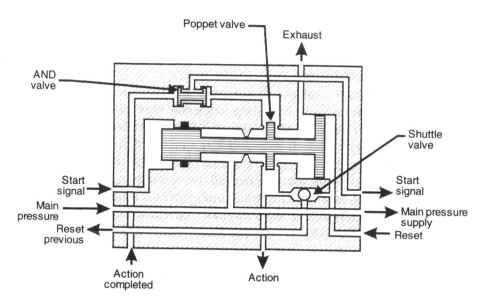

Fig. 14.14 Typical programme sequencer

14.10 Logic valves

Logic valves can be added to traditional circuits, as described in Chapter 12 and Chapter 16, Sections 16.1–16.3, to handle repeated or simultaneous inputs or sequences which would otherwise present difficulties. With stepper circuits as described in the previous section and Sections 16.5–16.7, logic valves are incorporated in the sequencer modules but are also frequently necessary to interface with the directional control valves which supply each actuator.

Logic valves were developed in conjunction with programme sequencers, so their configuration is non-traditional, symbols follow a different ISO standard, port numbering is different and mountings are designed for direct attachment to sequencers or logic subplates which are individual to each manufacturer. Logic or signal processing valves are often referred to as **gates**.

The **AND gate** is a three-port valve also named the **two-pressure valve** since it will only pass a signal to the output port if there is a signal on both input ports. Figure 14.15 shows the logic standard symbol, the alternative symbol which shows how the valve is actually constructed and the conventional symbol most closely representing the diagrammatic function of the valve. The 'truth tables' help to define the exact function. Note from the alternative symbol that one input only moves the poppet to seal the through path, but an input then added from the other side will pass through: also note that if the two input signals are unequal, then the lower pressure signal has priority.

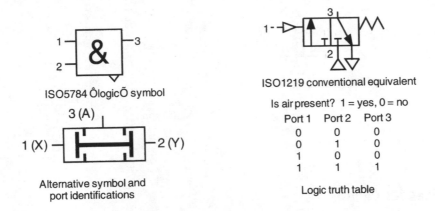

ISO5784 Ôlogic Õ symbol

3 (A)

1 (X) —| |— 2 (Y)

Alternative symbol and
port identifications

ISO1219 conventional equivalent

Is air present? 1 = yes, 0 = no

Port 1	Port 2	Port 3
0	0	0
0	1	0
1	0	0
1	1	1

Logic truth table

Fig. 14.15 The 'AND' gate

The **OR gate** is a three-port valve similar in function and internal design to the shuttle valve. A signal on either input port passes to the output port and results in the other input port being blocked. Figure 14.16 shows the logic standard symbol and the conventional ISO1219 symbol which represents the function of the valve. The 'truth tables' help to define the exact function.

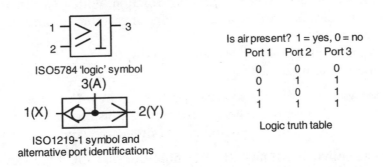

ISO5784 'logic' symbol

3(A)

1(X) —| |— 2(Y)

ISO1219-1 symbol and
alternative port identifications

Is air present? 1 = yes, 0 = no

Port 1	Port 2	Port 3
0	0	0
0	1	1
1	0	1
1	1	1

Logic truth table

Fig. 14.16 The 'OR' gate

The **NOT gate**, or **inhibition** valve, is a valve with pressure port, input signal port and output signal port, which will only deliver a signal to the output port if there is no signal to the input signal port. Figure 14.17 shows the logic standard symbol (easily confused with the & valve – note the input signal port has an 'O' added) and the conventional ISO1219 symbol which correctly represents the function of the valve. The 'truth tables' help to define the exact function.

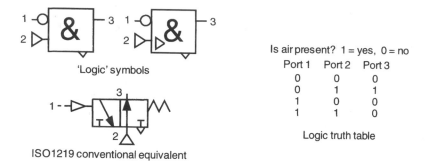

'Logic' symbols

ISO1219 conventional equivalent

Is air present? 1 = yes, 0 = no

Port 1	Port 2	Port 3
0	0	0
0	1	1
1	0	0
1	1	0

Logic truth table

Fig. 14.17 The 'NOT' gate

Exercises for Chapter 14

1. Give two applications for rodless cylinders and explain very briefly why conventional linear actuators are unsuitable. 2.5(a)

2. A mechanized paint spray powered by compressed air and controlled by a simple pneumatic control system comprises a vertical non-rotating pneumatic cylinder with the spray head fixed horizontally to the end of the actuator rod. The spray head is controlled to scan the entire work surface by regular up and down movements of the cylinder combined with regular turning of the entire assembly to left and right through 90° rotation. Which two of the following options would be effective and economical for turning the assembly:
(a) an air motor
(b) a double-acting cylinder actuating a crank arm
(c) a pneumatic rotary actuator. 2.5(a)

3. Outline two features of construction or function in which a diaphragm cylinder differs from a conventional linear actuator with piston. State two advantages of diaphragm type cylinders in comparison with piston type cylinders. 2.5(a)

4. Draw the conventional ISO1219-1 symbols for pilot/pilot 5/3 valves:
(a) with work ports pressurized with valve at rest;
(b) with work ports vented to exhaust when at rest;
(c) with all ports blocked when valve is at rest. 2.5(b)

5. For a particular circuit function, a double-acting cylinder must be served by a 5/3 directional control valve with work ports pressurized when the valve is at rest. Outline for each of the following types how the actuator will function in each of the three valve positions:
(a) a conventional double-acting cylinder
(b) a through-rod type of double-acting cylinder. 2.5(b)

6. A 5/2 valve and a 5/3 valve (whether solenoid or pilot operated) are similar in construction, but the 5/3 valve will have two more internal parts than a 5/2 valve. What are the extra components?   2.5(b)

7. Draw the diagrammatic circuit for a double-acting cylinder served by a 5/3 pilot/pilot directional valve, with a spring actuated cylinder locking device on the rod which is powered off by either pilot line to the DCV.   2.5(a)

8. For the AND logic valve, draw and give the port numbering for:
   (a) the diagrammatic circuit (ISO5784) 'logic' symbol;
   (b) the alternative symbol which illustrates the internal construction;
   (c) the diagrammatic ISO1219 symbol for the equivalent conventional valve but with 'logic' port numbers.   2.5(b)

9. For the OR logic valve, draw and give the port numbering for:
   (a) the diagrammatic circuit 'logic' (ISO5784) symbol;
   (b) the diagrammatic circuit ISO1219 symbol describing the function. 2.5(b)

10. For the NOT logic valve, draw and give the port numbering for:
    (a) the diagrammatic circuit 'logic' symbol;
    (b) the diagrammatic circuit ISO1219 symbol showing the function.   2.5(b)

15
Speed control and hydropneumatics

15.1 Metering-in and metering-out

To attempt precision control with pneumatics is like trying to tow a car with a rubber band. Small applied forces will change the volume of a column of air, so for precision applications incompressible fluids such as hydraulic fluid or even water are better. However, if a force is applied, a column of air at high pressure will change volume less than a column of air at low pressure. Speed control with air is therefore more resistant to force changes due to variable friction or loads if the restriction is applied as the air leaves the actuator than if the air is restricted on entry, as illustrated in Fig.15.1.

15.2 Hydropneumatic circuits

When used for a supporting function in a pneumatic circuit, a small hydraulic system comprising an actuator with a header tank and some flow control valves can be very compact and free of any regular maintenance requirements. By this means, precision speed control can easily be added to a pneumatic system. The pneumatics provide the force and the hydraulics provide the control, as shown in Fig. 15.2.

If on/off valves are substituted for the hydraulic flow restrictors, solenoid or air–pilot operated if required, then this circuit can lock rigidly in any required position. Alternatively, proportional electrically actuated hydraulic flow controls operating in a closed loop with a positional transducer on the actuator could provide precise control of speed and position, monitored and controlled by computer if required, all within a simple pneumatic machine.

The name given to add-on hydraulic speed control circuits is **hydrochecks**.

Figure 15.3 shows the **intensifier**, a way of generating high hydraulic pressures (raised in proportion to the areas of the cylinders) and also a local hydraulic supply, as required by some tool clamps for example. Reciprocating types are available, offering a continuous supply at up to 680 bar pressure!

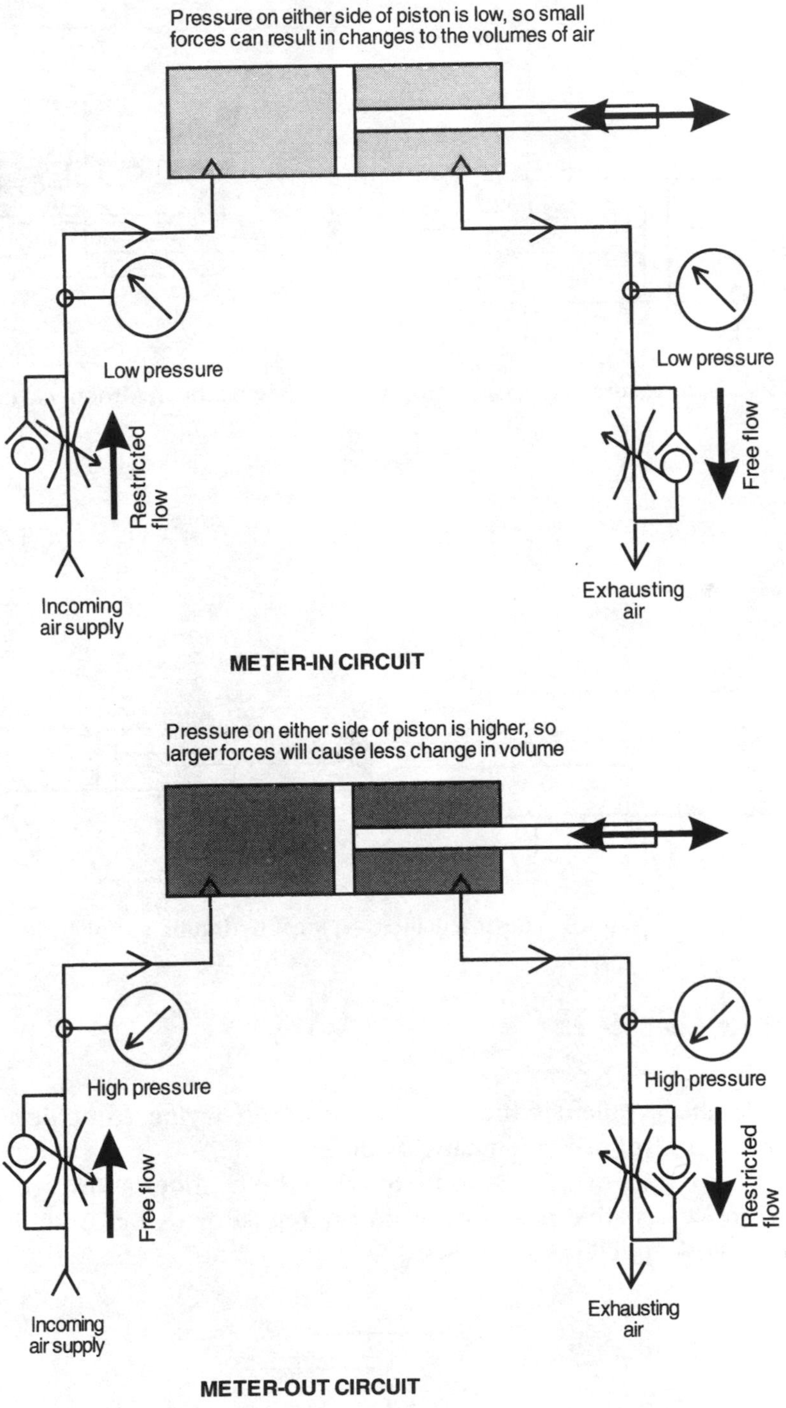

Pressure on either side of piston is low, so small
forces can result in changes to the volumes of air

Low pressure

Low pressure

Restricted
flow

Free flow

Incoming
air supply

Exhausting
air

METER-IN CIRCUIT

Pressure on either side of piston is higher, so
larger forces will cause less change in volume

High pressure

High pressure

Free flow

Restricted
flow

Incoming
air supply

Exhausting
air

METER-OUT CIRCUIT

Fig. 15.1 Control of actuator speed in air circuits

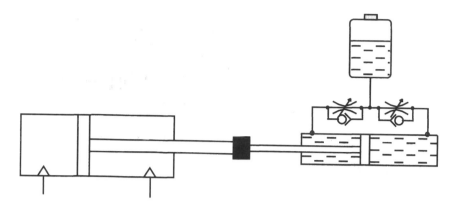

Fig. 15.2 Accurate speed control or locking with the 'hydrocheck' circuit

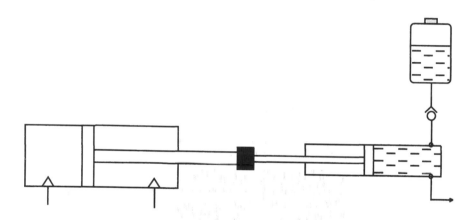

Fig. 15.3 The 'intensifier' – a local hydraulic supply

Exercises for Chapter 15

1. (a) State the problem which can occur when trying to achieve very slow speed control of a pneumatic cylinder.
   (b) A flow regulator valve is to be fitted to the cylinder shown below.
   Which alternative position for a flow regulator (A or B) will result in the best slow speed control?

2.5(a)

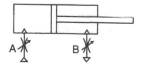

2. Draw the diagrammatic circuit for a 'hydrocheck' assembly, i.e. a pneumatic double-acting cylinder tandemed to a hydraulic double-acting cylinder, but in place of flow control valves to provide precision speed control, draw 2/2 on/off valves (solenoid/spring actuation) with reverse flow check valves, to provide rigid locking in any position.   2.5(d)

3. Explain in a few words the terms:
   (a) metering-in,
   (b) metering-out.   2.5(a)

4. A drilling machine powered by compressed air has a workpiece clamping arrangement incorporating miniature high pressure hydraulic actuator pads. Which of the following options would provide an effective and economical supply source for the local hydraulics:
   (a) a hydrocheck assembly for speed control using flow control valves;
   (b) an electrically powered hydraulic power pack with own reservoir;
   (c) a hydropneumatic intensifier assembly with single-stroke operation from a small header tank;
   (d) a hydrocheck assembly for rigid positional locking;
   (e) a hydropneumatic intensifier assembly with reciprocating operation providing a continuous supply.   2.5(d)

5. The adjustable jetting nozzle on a water turbine with a pneumatic control system requires rapid and extremely precise movement in response to signals provided by an electronic monitoring system. Which of the following options would be most likely to provide an effective and economical control system:
   (a) a hydrocheck assembly for speed control using flow control valves;
   (b) an electrically powered hydraulic power pack with own reservoir;
   (c) a hydropneumatic intensifier assembly with single-stroke operation from a small header tank;
   (d) a hydrocheck assembly for rigid positional locking;
   (e) a hydropneumatic intensifier assembly with reciprocating operation providing a continuous supply.   2.5(d)

16
Pneumatic circuits (3)

16.1 Cascade circuits

Chapter 12 introduced simple multicylinder circuits and the problems which arise due to 'maintained' and 'trapped' signals. Circuits to overcome these problems used pulsed signals or divided the circuit into zones into which the supply was latched by an additional directional control valve.

The **cascade system** was evolved as a simple method of designing more complex multicylinder circuits with up to four or more zones called **groups**.

The principles of cascade system circuit design are:

- A simple design procedure to determine the minimum necessary number of groups in the circuit to avoid trapped signals and then allocate the signal valves (i.e. limit switches) between these groups.
- Whereas multicylinder circuits in Chapter 12 have a single 'supply bus' carrying the main supply pressure to all components, cascade circuits have a separate supply bus for each group.
- A conventional arrangement of **selector valves** to divert the supply to whichever is the active group.

Applied carefully, the cascade system provides an easy and reliable method of specifying quite complex multi-actuator circuits which are often less costly than alternative systems because all the components are conventional.

16.2 Cascade circuit design procedure

The procedure for designing a cascade circuit is as follows:

1. *Write out the sequence.* A circuit should be regarded as operating continuously so the end of the sequence joins the beginning in a continuous loop. Reading from left to right, put a dividing mark before any single letter (neglecting

whether + or −) is repeated, treating the end as if it is joined to the beginning. The groups of functions between each division are the 'groups'. With some sequences, repeating the process reading from right to left results in fewer groups, treating the beginning as if it is joined to the end. The result with fewer groups is equally valid and will produce a simpler circuit.

As a very simple example, take the sequence from the circuit in Fig. 12.9,

i.e. A+ B+ B−A−

divided left to right = A+B+ / B− A− /
divided right to left = /A+ B+ / B− A− which (treated as a continuous
                                                                loop) is the same

numbering the groups: A+ B+/B− A−/
Group I   Group II

so this sequence can be delivered by **a two-group cascade circuit**.

2. *Draw the basic function diagram* (not including valve movements). Under *each* step in the function diagram, list *all* the limit-switch valves (assuming two per actuator) reading Chinese style from the top downwards as shown in Fig. 16.1.

3. No mental effort will now be required to draw a **grouping diagram** for the circuit.

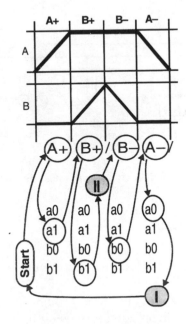

Fig. 16.1 The grouping diagram

Obviously, when actuator A extends (A+) it hits limit switch a1, and when B extends (B+) it hits b1. Step 1 on page 131 tells us that after B+ we switch to group II, so this change is inserted. When B retracts (B−) it hits b0. When A retracts (A−) it hits a0. After A− we switch to group 1.

4. *Draw arrows from step to step.* The design of this circuit is now complete.

It only remains to draw the circuit, as shown in Fig.16.2. The groups are supplied by a 5/2 directional valve, called a **group selector valve** in this application. To make the circuit as clear as possible, limit switches within the circuit are positioned near the DCV which they serve.

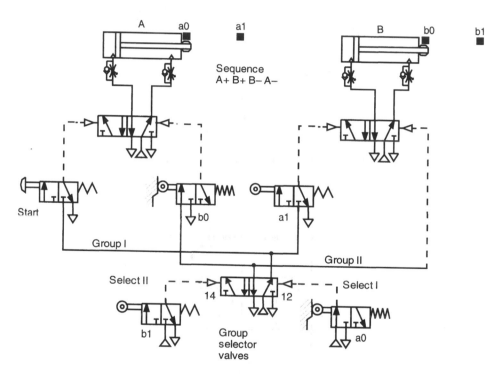

Fig. 16.2 A two-group cascade circuit

16.3 Practical cascade circuits

The same simple design procedure enables circuits to be devised to suit many practical applications.

More complex circuits may require more groups, but the design remains simple. In sample circuits showing linear actuators using limit switches for the signal elements, other forms of actuation and signalling, e.g. motors with pressure switches or timers, or rotary actuators with proximity switches, may be substituted as required.

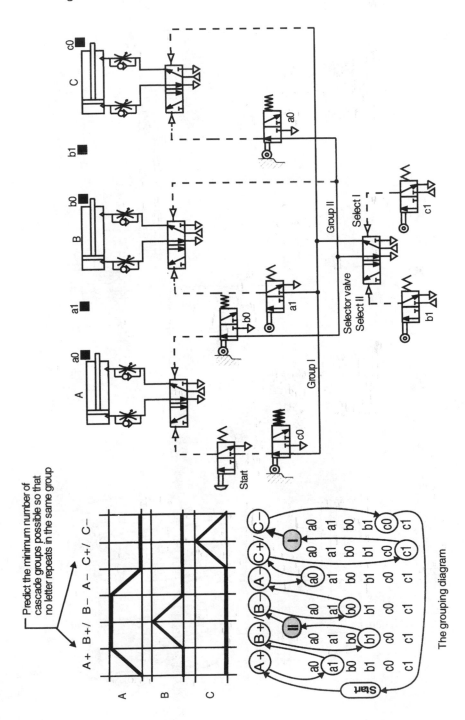

Fig. 16.3 A three-cylinder two-group cascade circuit

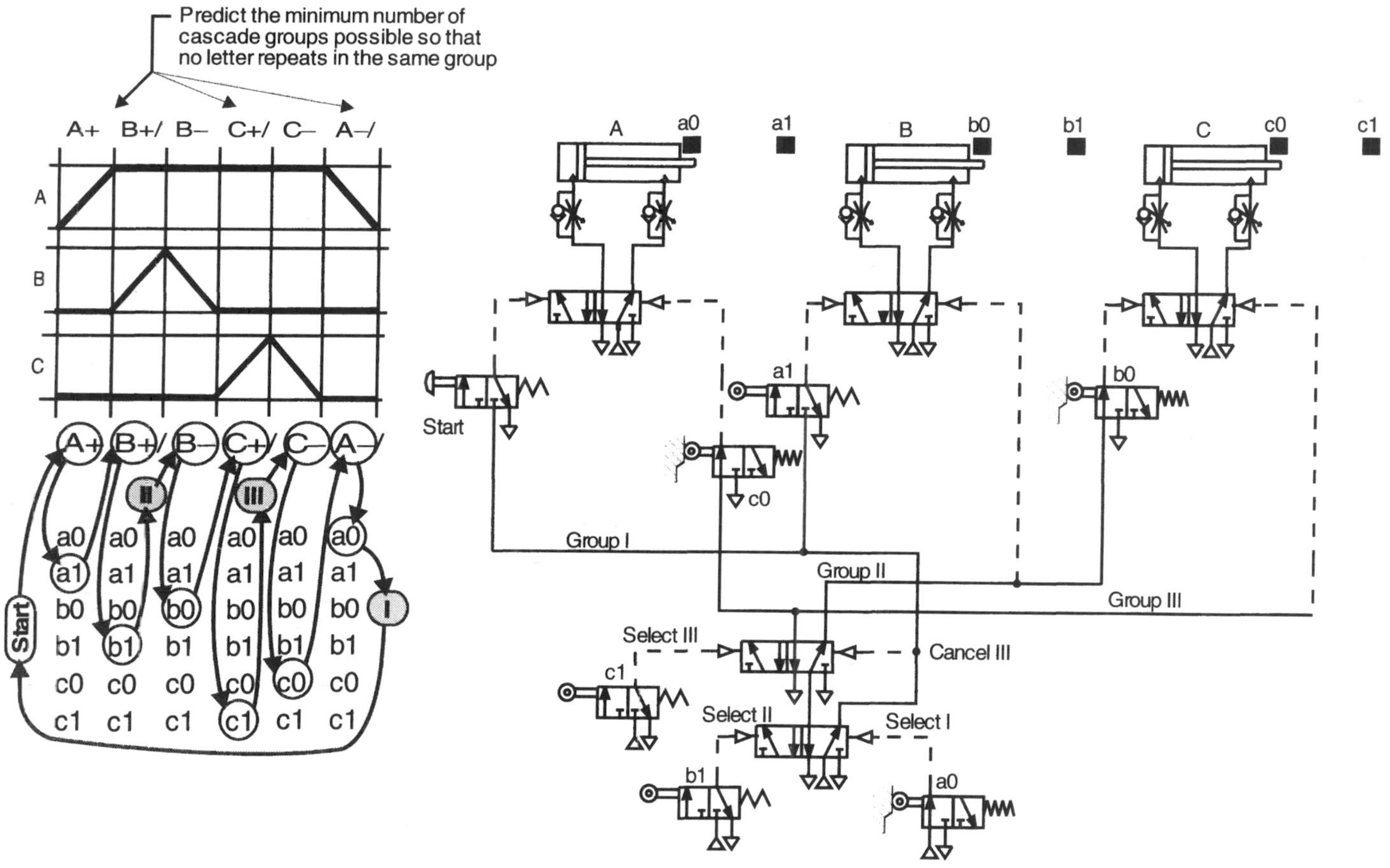

Fig. 16.4 A three-cylinder three-group cascade circuit

16.4 Group selection and stepper circuits

By determining the minimum number of supply groups for a cascade circuit, the circuit is kept simple and economical. The sample circuits in Figs 16.2, 16.3 and 16.4 use 5/2 pilot/pilot directional valves to supply each group. This results in the lowest possible component count and is shown again in Fig. 16.5.

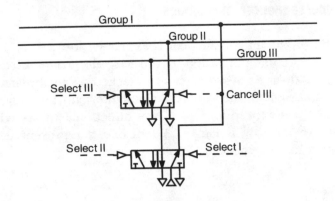

Fig. 16.5 Group selection with 5/2 valves

Figure 16.6 shows an alternative group selection circuit which requires more components but which is easier to follow. It is an equally valid way of supplying groups and it is often used. This is a classic **stepper circuit** because it steps through a predetermined sequence stage by stage and as each new step starts it includes arrangements to cancel the previous step.

Using cascade system technology to provide circuits using a minimum number of conventional components, each of which must be interconnected with pipework, is economical. However, modern design favours a modular approach in which components are manifolded together as far as possible, a consequence of the fact that parts are now cheaper than labour. The circuit in Fig. 16.6 conforms

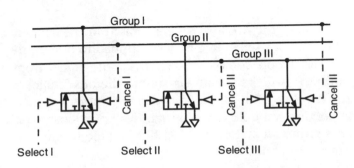

Fig. 16.6 Group selection with 3/2 valves

better with the modern design approach because it uses a succession of similar components with similar interconnections. It is relatively easy to design the circuit using 3/2 valves into a modular manifolded assembly, and a very small step from there to the programme sequencer described in Section 14.9, and shown in circuit form in Section 16.5.

16.5 Programme sequencer assemblies

The function of a programme sequencer is explained in Chapter 14, Section 9. It provides the same stepping function as the cascade group selection circuit in Fig.16.6 but in a modular form which can be stacked on a subplate, incorporating the cancellation and some of the signal elements internally, and requiring no external pipework except for the pilot legs to the directional valves which serve the actuators. Figure 16.7 below is Fig.14.13 repeated, a representation of the programme sequencer in conventional circuit format.

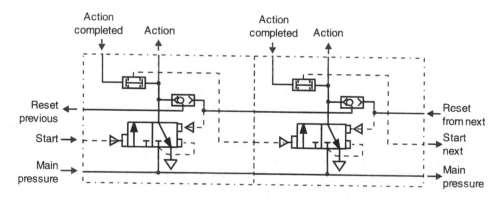

Fig. 16.7 Circuit for two programme sequencers

Note the function of the AND gate incorporated in the sequencer module: the action signal (i.e. the pilot signal to a DCV) is applied to one input port so there is no output until the action completed signal (i.e. the signal from a limit switch) arrives. This output provides the start signal into the next stage.

Note also the function of the shuttle valve in the sequencer module – it provides a signal to the previous module (and through to all previous modules in a multiple stack) if there is either an action signal in the module or any action signal in the next or subsequent modules.

Figures 16.8, 16.9 and 16.10 are slightly simplified sections through a triple stack of sequencers, showing how the internal spool/poppet valve and logic valves handle the stepping function.

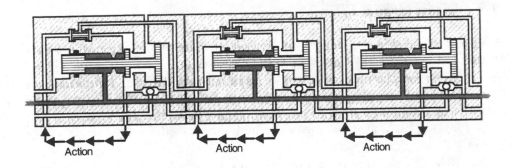

Fig. 16.8 Stacked sequencers with pressure applied and spools at rest – ready for start

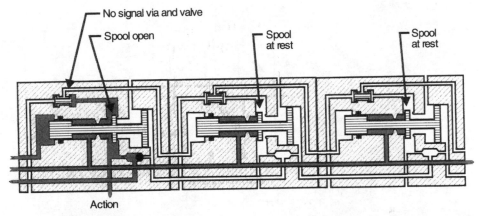

Fig. 16.9 Stacked sequencers with first start signal in, resulting in action signal out

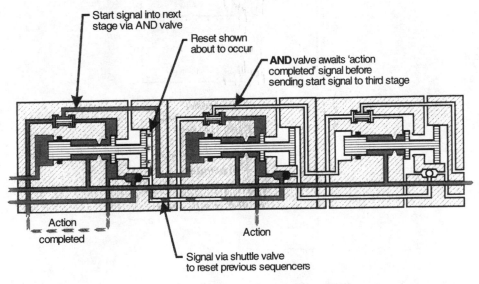

Fig. 16.10 Stacked sequencers with first action completed, resulting in start signal to second stage

16.6 Planning stepper circuits

Each programme step, as interpreted by a programme sequencer, comprises 'start', 'action' and 'verification' (meaning 'action completed'). An accepted format has been evolved for planning automation sequences so that the functions required can be translated very simply into a pneumatic circuit for a sequencer application, an electrical circuit for a stepper relay circuit (see Figs 17.8 and 17.9) or a program for a PLC application. This format, which originated in France, is called the **GRAFCET**.

The GRAFCET diagram is shown at its simplest level, called 'level 1' in Fig. 16.11 below; the sequence is described in words and laid out in a vertical table. A level 2 GRAFCET is shown in Fig.17.13, a PLC application of the same sequence. At

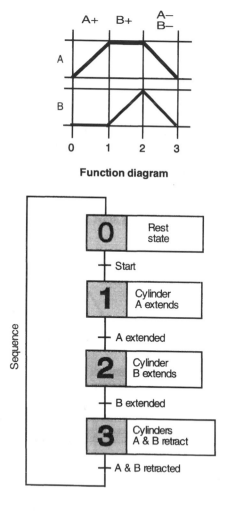

Fig. 16.11 The 'GRAFCET' diagram level 1 applied to a simple sequence (see also Fig. 17.13)

level 2, the words have simply been translated into the symbols we have already used, i.e. A+ B– etc. for actuator movements and a0 or b1 for the verification signals from limit switches for example.

The City & Guilds 2340 Part 2 Pneumatics syllabus only requires candidates to have outline familiarity with these concepts, since programming with particular application to PLCs is in the Part 3 syllabus. However, they are simple procedures evolved for use by technicians, commonly applied when PLC control sequences are planned, so many candidates will have the opportunity to see these procedures employed at their place of work.

16.7 Practical sequencer circuits

Figure 16.4 is a three-group cascade circuit with six steps. Figure 16.12 below shows a circuit giving the same sequence using a six-step programme sequencer assembly. If each sequencer module is counted separately then the sequencer circuit uses six more components than the corresponding cascade circuit. However, the design of sequencer circuits has a consistent and simple format, and trouble-shooting is easier because each step can be taken in isolation, so sequencers offer some real advantages for all-pneumatic circuits.

The step-by-step simplicity of the sequencer circuit makes it easier to accomplish simultaneous (two or more actuator movements occurring together) and also repeated movements by the addition of logic valves, as shown, for example, in Fig. 16.13. The OR valves (shuttles) on the output side of the sequencer feed signals simultaneously to two or more DCVs, and the AND valves on the input side make a step repeat until it has received two verification signals.

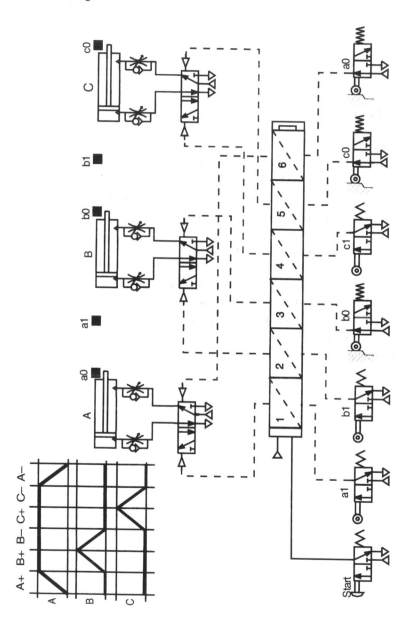

Fig. 16.12 Sequencer circuit for same sequence as cascade circuit of Fig.16.4

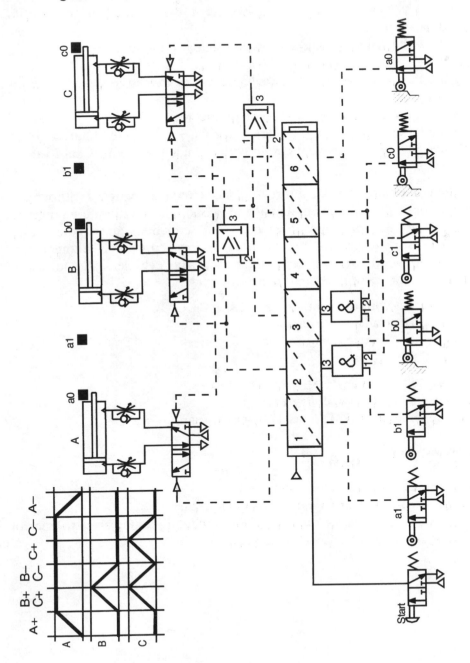

Fig. 16.13 Six-step sequencer circuit with simultaneous and repetitive movements

222 222 2222

Exercises for Chapter 16

1. For a three-cylinder circuit which: (i) raises a heap of waste material to a work area and holds it there, (ii) compresses it sideways then withdraws, (iii) compresses it lengthwise then withdraws, (iv) lowers it back to the position at which it started:
   (a) draw the function diagram showing actuator movements;
   (b) draw a GRAFSET (level 1) block diagram;
   (c) list all the limit switches in columns under each step of the function diagram;
   (d) work out the number of cascade groups required;
   (e) mark the sequence and groups by arrows joining listed limit switches;
   (f) design a cascade system circuit with a single press-button start valve.

   2.6(g)

2. Write down four alternative three-cylinder circuit sequences (without simultaneous or repetitive movements) and work out the minimum number of cascade groups applicable to each of your sequences. Show your sequences divided into the groups. Sketch the group selector valve arrangement that would apply to each sequence. Note that the cylinders should always be labelled A B C etc. *in order of their sequence.* 2.6(g)

3. For a system with the sequence A+ B+ A− C+ C− B− :
   (a) prepare a function diagram
   (b) plan the valve movements
   (c) design a circuit. 2.6(g)

4. For the sequence A+ B+ A− C+ C− C+ C− B−:
   (a) prepare a functional diagram
   (b) prepare a GRAFSET (level 1) block diagram. 2.6(i)

5. For the sequence $\frac{\text{A+ A−}}{\text{B+ B−}}$ B+ B−:

   (a) prepare a functional diagram
   (b) prepare a GRAFSET (level 1) block diagram
   (c) design a four-step sequencer circuit including logic valves to control the simultaneous and repeated movements. 2.6(f)

17
Electropneumatics

17.1 Solenoid valves

In the simplest electropneumatic circuits, the limit or proximity switch and valve solenoid connections are an electrical duplication of the pneumatic pilot circuits considered in preceding chapters. Figure 17.1 shows the diagrammatic representation of a linear actuator with proximity or limit switches together with the associated solenoid 5/2 directional valve.

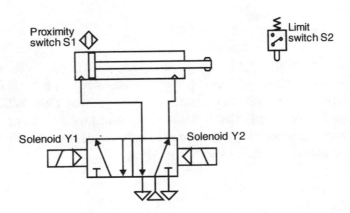

Fig. 17.1 Electropneumatic diagram basics

Figure 17.2 shows the construction of a solenoid spool valve, which is typically a pilot/pilot spool valve with the pilot supplies switched by small solenoid poppet valves mounted at the ends.

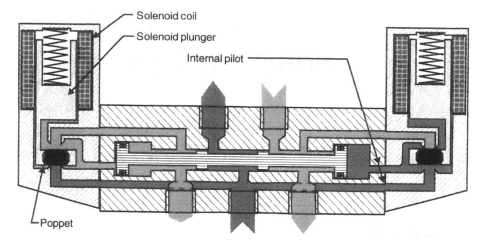

Fig. 17.2 Typical 5/2 solenoid spool valve

Usually solenoids have low current requirements that can be switched directly by a proximity switch, typically 3 W at 24 volts, equivalent to 125 mA, whereas a reed switch can handle up to 0.5 A or 500 mA. However, miniature reed switches and solid state proximity switches are often rated at 0.2 A. Wiring should normally include components to reduce 'back EMF', current induced by the solenoid coil as the signal breaks, which can damage the switches.

17.2 Pressure switches

A pressure switch (Fig. 17.3) changes over (opens or closes as required) at an adjustable pressure level, and can provide a signal to initiate a function in the sequence or perhaps simply to operate a visual warning light. When the pressure falls, the switch will change back. Usually for any given level of adjustment the switching pressure when the pressure is rising will be about 0.75 bar higher than the switching pressure when pressure is falling.

17.3 Sensors

Pneumatic machine functions frequently require the presence of a product or workpiece to provide a signal to progress the sequence. An example is where the intermittent arrival of a part-processed product from another machine must start an additional process in manufacture. Available types include purely pneumatic devices such as limit-switch valves or air catch sensors which respond to variations in back-pressure from a nozzle. In electropneumatic circuits, limit or proximity switches may be employed. Light-sensitive devices are also very reliable and particularly useful where products must be counted.

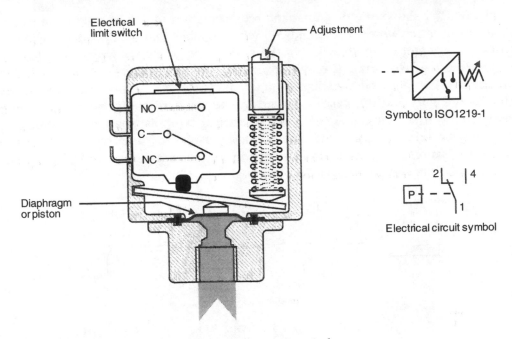

Fig. 17.3 Pressure switch

17.4 Relays

As the runner in a relay race passes the baton to a fresh competitor, so the electrical relay passes a weak signal to a stronger circuit. This can be necessary if the solenoid requires more current than the proximity switch can handle, as for example when several solenoids are switched together, or in order to use the capabilities of a number of relays to provide a stepping or logic function.

Figure 17.4 compares the single circuit for a directly switched solenoid with the two-part circuit for a solenoid switched via a relay. In multiple function circuit diagrams, the two circuits through the relay, the signal circuit and the actuator circuit, are drawn as separate circuit diagrams, the signal diagram and the actuator circuit diagram.

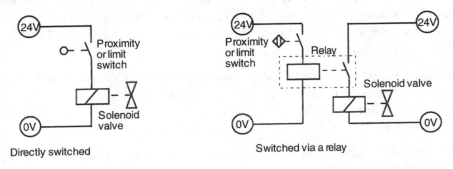

Fig. 17.4 Solenoid switching circuits

Figure 17.5 shows the diagrammatic representation of a 'four pole changeover relay', or in other words, a solenoid coil which pushes four independent switches from one contact to another within the same package. Relays with multiple contacts are often used even if only one or two sets of contacts are required. The apparently complicated two-digit contact numbers are simply the contact set (numbered 1 to 4 in this case) followed by the standard contact number for the switch type fitted, as explained in the 'electrical symbols' chart in Appendix C. Latching or 'bistable' relays are also common, using 'set' and 'reset' coils. These retain the switched position after the energizing current is removed which gives simpler circuits and advantages for some logic functions.

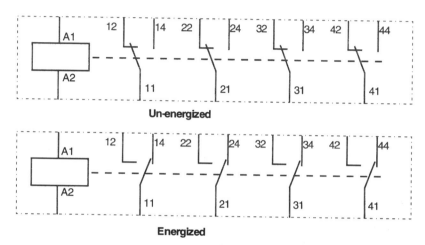

Fig.17.5 A typical multicontact relay

17.5 Relay control

Simple electropneumatic circuits which electrically duplicate the functions of simple purely pneumatic circuits experience the same problems due to trapped and maintained signals when more ambitious sequences are attempted. To avoid signal conflicts, relays can be used to step through the sequence similarly to the pneumatic sequencer (see Section 16.5). Figure 17.6 shows the simplest way this can be done, using latching type relays. (Numbering of relay contacts is explained in 'electrical symbols', Appendix C). Each grey area is a step, started by the proximity switch (S1, S2 etc.) closing, which:

- energizes the appropriate solenoid and therefore starts the actuator;
- resets the previous relay to stop the previous actuator;
- introduces a supply to the next proximity switch which will make connection when the actuator stroke completes and therefore start the next step.

For a stepping circuit using standard ('monostable') relays which return to the normal mode when the current is removed, a more complex circuit is required, as shown in Fig. 17.7. Each relay coil is supplied via a normally closed contact in the

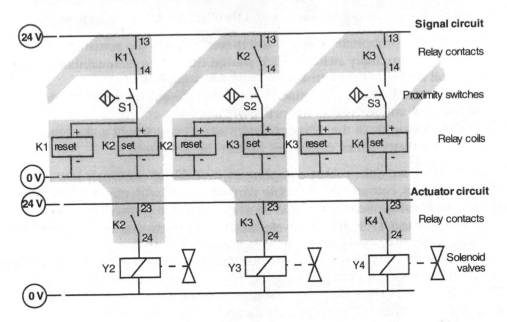

Fig. 17.6 Latching relay stepping circuit

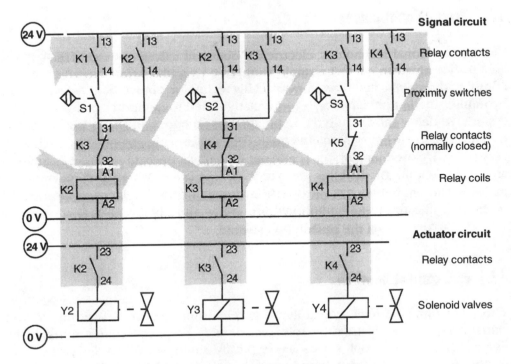

Fig. 17.7 Standard relay stepping circuit

next relay. As each relay coil is energized through the stepping sequence it:

* energizes the appropriate solenoid and therefore starts the actuator;
* bypasses the proximity switch which started the step – this maintains current to the coil as the actuator stroke continues to completion;
* opens the normally closed contacts which completed the supply to the previous relay coil, and so cancels the previous step;
* introduces a supply to the next proximity switch which will make connection when the actuator stroke completes to start the next step.

Miniature relays are quite robust components, available at extremely low cost. For training purposes, colour coded flexible wiring secured with clips may be acceptable, but practical applications should be constructed on printed circuits.

Figures 17.8 and 17.9 are simple electropneumatic circuits for a two-actuator sequence for which a pneumatic circuit is shown in Fig.16.2.

The circuit in Fig.17.8 uses four 'monostable' (standard) relays, each with three sets of changeover contacts. The stepping function is illustrated in Fig.17.7.

The circuit in Fig.17.9 uses four 'bistable' (latching) relays, each with three sets of changeover contacts. The stepping function is illustrated in Fig.17.6.

By changing the designations of proximity switches and solenoids, plus more actuators if required (each controlled by two relays), these circuits can be adapted to suit many applications.

17.6 Proportional valves

Using proportional technology, electrically operated valves can vary flow. A variable orifice valve can control a single supply, or a variable stroke spool valve can combine directional and flow control. This technology, much better developed in hydraulics than in pneumatics, is particularly suited to computer control, enabling systems to deliver any required combination of functions and speed. By accurately measuring actuator movement and passing this information electrically back to the controller, a 'closed loop system' is created which can provide precise positioning anywhere within the working stroke of an actuator. Appendix C 'Symbols 2' includes the symbol for a 3/3 proportional directional valve. The proportional features are indicated by the additional lines above and below the boxes, together with the arrow through the push–pull solenoid.

17.7 PLC control in outline

The programmable logic controller is an electronic device delivering electrical signals to drive pneumatic valve solenoids directly. No pilot system is normally necessary. Directional spool valves are typically multiple-banked in 'islands' with multiwire connection (many conductors in a single cable) or field-bus connection

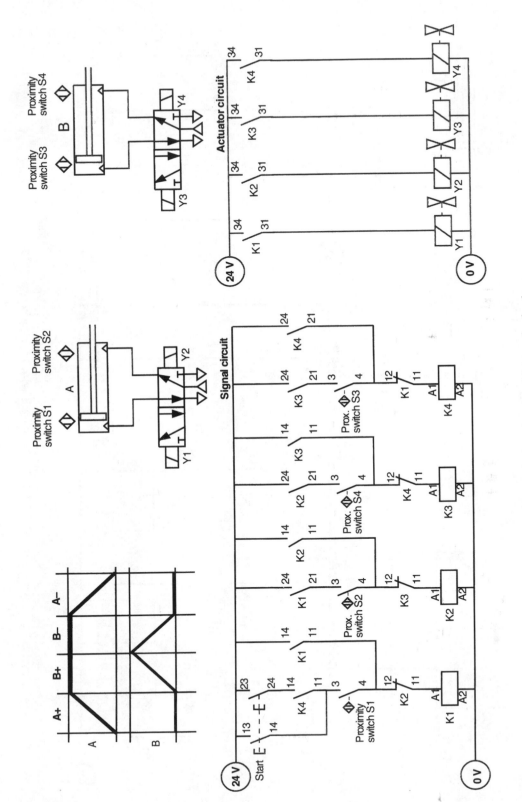

Fig. 17.8 A relay circuit with standard relays

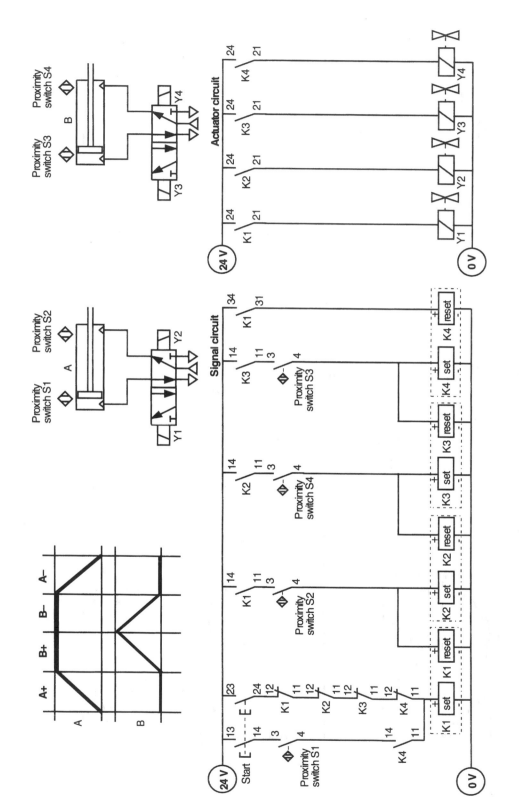

Fig. 17.9 A relay circuit with latching relays

(coded signals via two or more conductors in a shielded cable) between the islands and the PLC. Signals from the PLC, which includes timers, counters and shift registers, can be arranged to give any required sequence and some PLCs also provide 'analogue' (i.e. variable) signals for proportional flow devices. The PLC sequence must first be programmed into it using a plug-in programmer or alternatively a computer.

To introduce this technology, methods of operation were developed which are similar to methods employed previously, so PLC programs are often planned by constructing electrical circuit diagrams using the US symbol standard and diagram arrangement, called 'ladder diagram format' in this book to distinguish it from the DIN40713 standard previously explained. If a computer is employed for the programming, there may also be the option of using the GRAFCET format as for pneumatic sequencers. The program instruction codes include familiar pneumatic logic terminology such as AND, OR and NOT. Figure 17.10 illustrates a PLC controlled pneumatic system excluding the actuators.

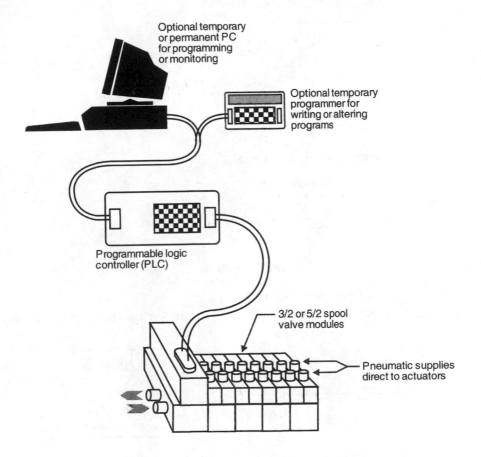

Fig. 17.10 PLC control technology

17.8 Ladder diagrams for sequence and PLC control

Although DC electrics behave much like air circuits, pneumatic logic circuits cannot be translated into either electrical circuit format by a process of simple substitution. The electrical circuit equivalent to a pneumatic circuit gives the same result but by different methods. Figure 17.11 below shows how a pneumatic circuit which requires AND, OR and NOT logic gates has a function which can be represented by an electrical diagram featuring only NOT and OR gates. Figure 17.12 shows the electrical diagram AND function.

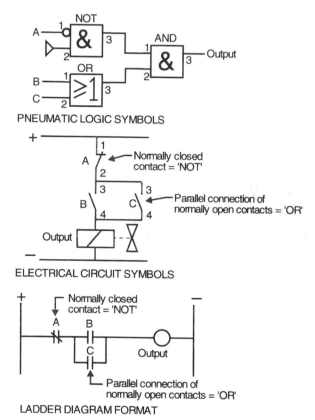

Fig. 17.11 Logic circuits – for an output, A must be 'off' and B or C must be 'on'

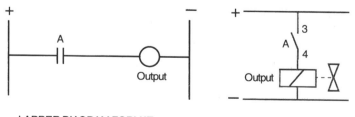

Fig. 17.12 The 'AND' function – for an output, A plus supply must be 'on'

Figure 17.13 shows a two-cylinder circuit with actuators and directional control valves (which can equally represent banked spool valves), the GRAFCET, the corresponding ladder diagram and a representative set of PLC instructions. See also Fig. 16.11 for the GRAFCET level 1 for the same sequence.

The PLC instructions (which vary according to manufacturer) are abbreviations for programming words as follows: FUN, function; ORG, origin; STR store; OUT, output.

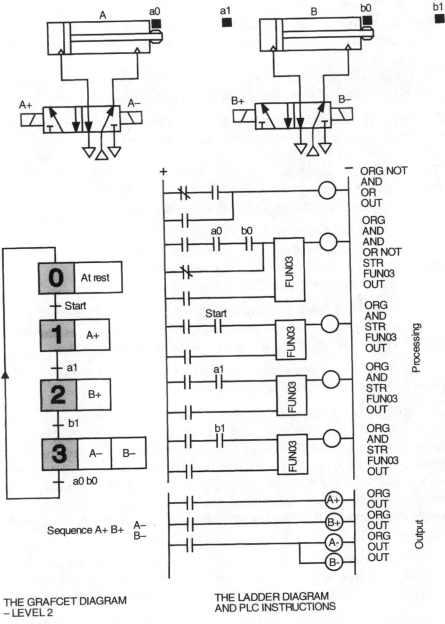

Fig. 17.13 Stages in PLC programming

Exercises for Chapter 17

1. A simple single actuator circuit employs a large solenoid valve requiring an operating current of 0.3 A which must be switched by a miniature reed switch with a current capacity of only 0.2 A. What single component must be introduced into the circuit? Briefly explain the effect of the additional component.
   2.6(i)

2. Draw the conventional ISO1219-1 symbol for:
   (a) a solenoid operated 5/3 valve with work ports vented to exhaust when at rest;
   (b) a solenoid operated proportionally controlled 3/3 valve with push-pull solenoid and all ports closed in the at-rest position.   2.6(b)

3. For each of the following components sketch the diagrammatic symbol employed in (i) a pneumatic circuit, (ii) an electrical circuit:
   (a) proximity switch
   (b) electrical limit switch
   (c) pressure switch
   (d) direct solenoid actuation of a directional valve.   2.6(a)

4. For the sequence A+ B+ A− C+ C− C+ C− B− :
   (a) prepare a functional diagram
   (b) prepare a GRAFSET (level 1) block diagram
   (c) prepare a GRAFSET (level 2) block diagram.   2.6(i)

5. For each of the following, draw a single line of circuit diagram (using US ladder diagram format symbols) showing + and − bus lines, gates and an output:
   (a) the NOT function, (b) the OR function, (c) the AND function.   2.6(i)

6. Explain briefly the essential difference between a monostable (standard type) relay and a bistable (latching type) relay.   2.6(i)

7. In relay stepper circuits, using either type of relay, each step is started by a proximity or limit switch closing to verify that the preceding step is complete. However, although this switch may then open, the relay contacts retain position until the next proximity or limit switch closes to signal that the current stroke of the sequence is completed. Explain briefly with the help of a simple sketch, the method used to maintain the relay position, for (a) a circuit employing bistable (latching) relays; (b) a circuit employing monostable (standard) relays.   2.6(i)

8. Draw the electrical circuit symbol for a multicontact monostable (standard) type relay with one coil and four sets of contacts each of which change from one independent contact to another when the coil is energized.   2.6(a)

9. Sketch the diagrammatic electrical circuit symbol for the following types of switch and indicate the single digit contact numbers:
   (a) normally open proximity switch
   (b) push-button switch with changeover contacts. 2.6(a)

18
Maintenance

18.1 Types of maintenance

Maintenance repairs are carried out in response to problems. An outline list of typical problems with typical causes is given in Section 5.6. **Routine maintenance** is compliance with the servicing requirements of equipment at the intervals specified by manufacturers. This also is outlined further in Section 5.6.

Preventative maintenance is commonly embodied in a planned preventative maintenance (PPM) system – a schedule of inspections and good housekeeping practices designed to avoid the occurrence of problems, principally based on experience.

When maintenance repairs occur, the maintenance staff are the victims of events, but a good PPM system puts the staff in control. In an ideal plant the routine maintenance and planned preventative maintenance would be so comprehensive and effective that repairs would never be needed. This may be impossible to achieve, but if a maintenance manager wants maximum availability of plant delivered by the smallest possible staff then attention to the PPM system will give the best results.

18.2 The planned preventative maintenance system

Manufacturers of primary plant equipment point the way to an effective PPM system. A typical compressor, for example, has scheduled maintenance operations to be carried out after specified periods which are established with regard, for example, to knowledge of the length of life of the lubricating oil, or the typical rates of contamination of filter elements. This routine maintenance on specific primary items of plant is complementary to the PPM system, which is an extension of routine maintenance applicable to all the compressed air and pneumatic equipment and services on the site.

The PPM system can cover **replacements**, **adjustments** and **checking**.

In the PPM system, replacements are made not because failure has occurred, but according to a schedule for replacement of the part before its useful life is over. Adjustments are made not because the performance of the equipment is unacceptable but to a schedule drawn up to anticipate the effects of wear or usage. Checking is necessary when knowledge of all the things that can go wrong with a plant is incomplete. Well-kept records of the results will mean less and less checking having to be carried out.

For replacements and adjustments, the way in which manufacturers specify maintenance periods can be extended to the whole plant. For all machines or components that are not part of the primary manufactured equipment (i.e. not already subject to routine maintenance recommendations), each item can be given a 'maintenance life'.

Each element could be considered according to the following criteria:

(a) degradation of material or wear due to use
(b) biological and physical degradation of material
(c) degradation of material or wear due to abuse
(d) replenishment requirements due to losses
(e) replenishment requirements due to usage
(f) adjustments to compensate for wear
(g) seasonal component replacements or adjustments
(h) component replacements due to contamination.

For example, applying the above headings to a portable drill, we obtain the following table of life (hours):

Component	Bearings	Gears	Chuck	Governor	Valve	Hose	Seals
(a)	10 000	7500	2500	7500	5000	2500	None
(b)	None	None	None	10 000	5000	5000	5000
(c)	5000	6000	1000	None	None	1000	None
(d) (e) (f) and (g)			Not applicable				
(h)	None	None	1000	2500	None	None	None
Scheduled life:	5000	6000	1000	2500	5000	1000	5000

Applying the headings to a multiple wedge-belt drive, we obtain:

Component	Bearings	Pulleys	Hub-bush	Belts
(a)	10 000	5000	None	2500
(b)	None	None	None	5000
(c)	None	None	5000	None
(d) and (e)		Not applicable		
(f)	None	None	None	1000
(g) and (h)		Not applicable		
Scheduled life	10 000	5000	5000	1000

These schedules would determine replacement and adjustment periods together with stocks of spare parts to be carried, and show how a PPM system to provide replacement and adjustment intervals for all the components in the plant may be devised from scratch.

However, the more important input in the longer term is records of **actual** service times and problems encountered.

If service records are maintained to give regular inputs for the purpose of revising the above tables of life for the plant components, then an excellent PPM system tailored to the real needs of the organization will evolve.

Examples of checking within a PPM system would be:

* regular checks under varying conditions of demand to ensure that the compressed air main remains at required pressure and delivery;
* the checking of individual tools for handling and manipulative damage;
* checking during and after use to ensure that tools are properly handled, cleaned and correctly stored;
* the checking of operator working practices together with a programme of operator instruction;
* checking the speeds of rotary tools and drills with a tachometer;
* checking the working temperatures of machinery and flows of coolant to ensure that they are adequate under all conditions;
* checking of shutdown and fail-safe devices by simulating plant failure or incorrect use;
* regular surveys to monitor excessive machinery noise or evidence of air leakage.

18.3 Legal requirements

The Pressure Systems and Transportable Gas Regulations 1989 include requirements affecting maintenance operations and procedures. These are outlined in Chapter 8. Note also that under the regulations:

Periodic examinations must be carried out using a written scheme of examination, prepared and carried out by an appointed 'competent person', resulting in a written report. The report will note the condition of the parts of the system examined and if necessary specify any recommended repairs, modifications or changes together with dates by which they must be implemented, plus a further examination date to be marked on the equipment, after which it must not be operated. It is a responsibility of the competent person to advise the user if there is 'imminent danger', in which case the equipment can be completely or partially closed down until the recommended work has been carried out and examined.

Records must be kept of:

* the latest periodic report as outlined above
* previous reports if relevant to safety assessments
* details of design and construction

- rated conditions of operation
- details of all repairs and modifications
- all foreseeable faults together with the operation of safety devices.

18.4 Safety of personnel

Maintaining pneumatic machines can be so dangerous that all factories have carefully devised safety procedures for maintenance staff which must always be given precedence over the demands of production.

A typical procedure for ensuring safety when dealing with system breakdowns or component failures would be as follows. At start of the repair operation:

- ensure that suitable lifting and handling equipment is to hand;
- obtain full details of the fault as understood by the operators (see also the procedures for maintenance repairs in Ch. 19 under fault-finding) and note the positions of all actuators and valves;
- all air pressure should be completely isolated and vented from the system together with all electrical supplies except in the case of 'inspections under working conditions';
- ensure no loads remain applied to the system actuators;
- protect against inadvertent starting of the appliance by Permit to Work procedure or applying labels to controls etc.;
- removal of any component, pipework or product from the system should be undertaken with precautions to allow for the effect of trapped air under pressure;
- if checking actuator alignment, whether under pressure or not, ensure that fingers or other parts cannot be trapped;
- check pressure rating of all replacement parts fitted.

On restarting system on conclusion of repair:

- ensure that air connections are mechanically tight and secured;
- double check that pipe connections are made to the correct ports;
- ensure that all electrical devices are properly installed, suitably fused, insulated and earthed;
- before start-up ensure all actuators are in correct 'at rest' positions;
- when starting or adjusting a system, take care that fittings do not blow out under pressure;
- always run a system at the lowest practicable pressure;
- silencers, exhaust port filters and other types of port fittings should not be removed whilst the system is pressurized;
- never subject the body to compressed air by misapplication of jets of air or by attempting to block exhausting orifices;
- pressure should be introduced slowly via manual pressure regulator or by effective soft-start device;

- check the operation under fully safe conditions, i.e. with guards or other safety arrangements in operation.

Exercises for Chapter 18

1. Explain very briefly the essential difference between:
   (a) maintenance repairs
   (b) routine maintenance
   (c) preventative maintenance. 2.9(a)

2. Give six examples of routine checking operations as carried out within a plant preventative maintenance system (PPM). 2.9(a)

3. State six of the precautions for ensuring safety which should be observed when:
   (a) starting maintenance/repair of a plant
   (b) completing overhaul, before and during start-up of the plant. 2.9(c)

4. Name four checks you would expect to make for regular maintenance of the air treatment installation shown diagramatically below:

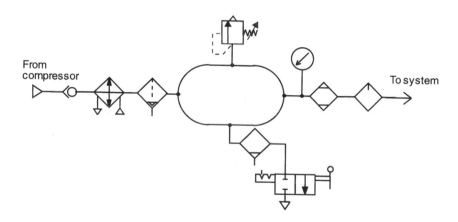

2.9(a)

5. Briefly describe five of the records required to be kept after a periodic examination following repairs and modifications of a compressed air installation which have been carried out to the requirements of the Pressure Systems and Transportable Gas Regulations 1989. 2.9(d)

19
Fault-finding

19.1 The approach to fault-finding

Section 5.7 provides a useful summary of the approach to fault-finding, which is covered in this chapter with some more practical information and guidance. A main objective for students of pneumatics is to test and improve their competence in the diagnosis and efficient repair of faults. Candidates often approach fault-finding with such confidence that they see no need for method or documentation and no need sometimes even for information describing the functions of a circuit. Time always shows this to be unjustified: it can soon be demonstrated that all faults can be traced much more quickly using the **diagnostic approach**. Furthermore, a 'pass' cannot be achieved in the assessment of competence unless diagnostic methods have been used and adequately recorded in the report. These comprise the following:

- The use of available records – an up-to-date circuit diagram at the very least, and attention to reports of previous repairs. Attention should also be given to pipework colour codes which should also be marked on circuits.
- Understanding the functions of the system as indicated by the sequence of movements of actuators, plus the valve positions through the sequence as indicated on a full function diagram if available.
- Identification of the complete symptoms of the fault – do not imagine that production personnel will always give you this information, or, if they do, that it will be accurate!
- With reference to the circuit diagram, list possible causes of the fault.
- Systematically investigate non-pneumatic reasons, mechanical defects or operational abuse, in particular why the fault could occur.
- If the system stopped in mid-sequence, prepare a list of all actuator and valve positions. Comparing this list with the positions shown on the full function diagram will often give an immediate indication of the fault.
- Before starting work on the hardware, systematically look for any safety risks resulting from the failure or from the fault-finding procedure.

- Carry out the investigation of the condition of the hardware by following a numbered progression through the circuit, identifying each step on the circuit diagram. Figure 19.1 shows an example of this, in which each step is identified by the component designation and port number.
- Always leave the installation in a better condition than it was previously, by furnishing brief and well-presented documentation to assist troubleshooting in future. For example, if a full function diagram is not provided, prepare one. If pipework is not colour coded, ensure that this is done.
- As the investigation proceeds, note each step taken, together with the results, in a report. Note also the legal obligation for reports of repairs (see Section 18.3).

Throughout an investigation, repair and recommissioning, safety procedures outlined in Sections 5.4–5.6 and 18.4 should be given overriding priority.

19.2 Common faults

The faults described in Section 5.6 are discussed below, with more comprehensive notes on typical causes.

Excessive noise   This may be pneumatic noise originating in inefficient or defective components and transmitted by the air as pulsations. Its creation should be prevented if possible, bearing in mind that some noise is inevitable, and then the transmission of noise should be approached separately. A simple stethoscope (or just a rod to the ear applied to components or pipework) enables noise to be tracked through a system and can help to distinguish between pneumatic and mechanical noise. Compressors are the most likely source, with excessive levels sometimes resulting from worn drive couplings, bearings or piston rings, or from an accumulation of water or sludge reducing the cylinder clearance volume. Noise transmission can be reduced by resilient pipe supports, which if not available from compressed air equipment suppliers, can be bought from hydraulic distributors for whom noise is a more serious and common problem.

Vibration   Vibration may be due to low frequency noise created by a worn compressor. High pitched vibration can be evidence that a relief valve is venting. Slack fastenings in a bolted framework can reveal vibration of pneumatic or mechanical origin and may be potentially dangerous.

High temperatures   High temperatures in distribution indicate ineffective cooling. Small quantities of sand or solids deposited in a heat exchanger water jacket should be suspected, or coolant pump drive belts may need adjustment. High actuator or exhaust temperatures are a sign of friction caused by contamination or poor lubrication, a worn or faulty motor or perhaps a high side load on an unsupported actuator rod resulting in worn bearings.

Contamination   Contamination means that filters must be checked and cleaned or replaced, and then the source of contamination must be traced. Water and sludge can be evidence of a defective dryer or badly designed distribution system with water pockets. Dirt or gummy deposits may indicate insufficient lubrication or use of the wrong oil. Alternatively, seals may be disintegrating due to old age, poor lubrication, high temperature or contact with aggressive fluids.

Erratic operation   This usually means that air pressure is low in relation to the applied loads. Common causes are increased demand from other users resulting in an air shortage, an increase in applied load due to changed usage, or mechanical friction resulting from worn parts or bad lubrication. An alternative cause is incorrect adjustment or settings of speed controls. Smooth operation is achieved by using the highest air pressure possible at the actuator, provided a safety margin is allowed for load variations. Pressure is normally adjustable by 'metering out', i.e. by means of speed control valves in the air exhaust path. Purely mechanical reasons for erratic operation should also be checked before adjusting the air system.

Another reason for erratic operation, which can be difficult to diagnose, is the minimum working pressures applicable to some small poppet valves, particularly limit-switch valves, which may need about 3.5 bar before they operate correctly.

Leakage   Leakage may be internal, causing reduced or uneven speed of operation, possibly accompanied by reduced available force from actuators or motors. The condition of seals and internal components should be inspected. External leakage may have the same symptoms but also results in noise, and in minor instances can be detected with soapy water. Pipe connections are usually responsible for external leaks. Screwed joints may need reassembly with the use of sealants, or new olives may be needed for compression fittings. Small amounts of leakage are widely tolerated in pneumatic systems with no ill effects whatsoever until the total demand for air exceeds the supply.

Excessive pressures   These cause a system to stop due to overload so that the regulated supply pressure must be raised. In extreme cases, the rated capability of the compressor and air receiver are reached, so that venting of the relief valve occurs. Pressures only rise due to increased loads or unsuitable adjustments of speed controls. Loads, however, can rise due to mechanical factors such as friction due to poor lubrication, worn bearings and bushes.

Incorrect speed of operation   Incorrect speed of operation usually follows increased demand elsewhere in a distribution system leading to an air shortage. Leaks, particularly internal leaks in an actuator or motor causing slippage due to worn components or seals, can result in loss of speed. Some air tools incorporate governors to limit speed, which can stick or become sluggish in action due to wear, with the result that speeds can rise.

Incorrect sequence of operations   This is an uncommon fault caused by loose or defective proof-of-position valves, tampering with or poor electrical installation of

PLCs, loose connections to solenoids etc. Some circuits which rely on pulsed signals, often in association with counters, can be prone to irregular operation which may be cured by changing the duration of the signals.

Hose, tube and pipe failures   These usually result from vibration and inadequate support. Plastic hose materials are generally unsuitable for temperatures over 60°C or below freezing. ABS (usable up to 70°C) and PVC (usable up to 50°C) 'rigid' materials for distribution pipework require support at between 15 to 40 pipe diameters (depending on size) at 20°C, but at only 8 to 24 diameters at 50°C. Allowance must also be made for expansion and contraction with all pipe materials by the use of loose-fitting clips and the incorporation of expansion loops in long runs. Heavy components should never be hung on the pipework alone, but should be provided with additional mechanical support.

19.3 The use of test equipment

Fault-finding can often be carried out without test equipment, since a pressure gauge is normally to hand in the local service unit and the main indicator for fault-finding is the presence of air at a given position in the circuit. However, the only safe way to test for air requires the circuit to be shut down, the pipework broken at the test position and the circuit started again.

Alternatively, a number of pressure gauges with tee-pieces can be inserted into the 'dead' circuit so that when started there is an instant view of the local pressure of air at a number of locations. 'Pneumatic indicators', compact panel fittings which change colour when pressure is present, can be assembled into a 'diagnostic panel' showing the locations on the circuit where there is pressure. 'Self-seal' push-in tube couplings are also available which automatically close when a tube is removed, and which are thus useful as pressure gauge test points. If troubleshooting is expected, these facilities save time and improve safety.

Flow meters and tachometers are required mainly for work on compressors and motors. The typical flow meter is a glass cylindrical coupling for insertion into the pipeline with a moving piston to indicate the flow in litres/min. This enables the output of a compressor to be checked, changed by calculation into free air and compared with the rated performance. The driven speed of the compressor is that of the driving motor, but if this is in doubt, a tachometer enables it to be checked. The typical tachometer is a battery powered hand-held torch-like instrument with a digital display and photoelectric send–receive device which when pointed at a reflective spot (tape fixed beforehand or a painted spot) on shaft or coupling gives an accurate read-out in rpm. By this process, together with a pressure gauge in the delivery line, it is also possible to test a motor running free or perhaps working under load.

19.4 Functional charts for fault-finding

Functional charts which show all the valve movements can save hours of work if they are available for troubleshooting. Consider a system such as that in Fig. 19.1 which has stopped working at the stage indicated by the grey line. All the actuator and valve positions can be seen at a glance. Furthermore, for grouped, cascade or latching circuits, if the diagram is drawn to indicate which sections of the circuit are active (as shown by solid or dotted lines), testing and analysis of the circuit are simplified. The system here is a very simple one (see Fig.12.9), but with more actuators and valves the advantages are greater. Comparison of the actual system state with the diagrammatic representation will often give immediate indication of the fault.

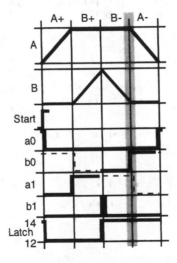

Fig. 19.1 Functional chart or diagram

19.5 FCR procedures and algorithm charts

A popular and effective method for fault-finding is the fault cause remedy (FCR) procedure – a structured route to identifying and fixing faults.

Algorithm charts are often found in manufacturers' equipment manuals – a progressive list of possible faults in a question and answer routine giving alternative routes to a solution.

Figure.19.2 shows an algorithm chart for the circuit illustrated which is based on the fault cause remedy procedure.

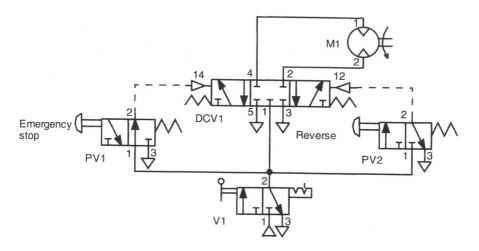

Step No.	Fault?	Y/N?	To determine cause	Cause identified?	Remedy	Cured Y/N?
1 Check supply	Air at V1-1?	Y N ↓ 14	→ Is isolator valve open?	N Y ↓ 2	→ Open valve	Y N ↓ ↓ 14 2
2 Check regulator	Air at V1-1?	Y N ↓ 14	→ Regulator correctly adjusted?	N Y ↓ 3	→ Reset regulator	Y N ↓ ↓ 14 3
3 Check soft start	Air at V1-1?	Y N ↓ 14	→ Test valve functions and solenoid control if fitted.	N Y ↓ 4	→ Replace valve	Y ↓ 14
4 Check exhausts PV1-3 PV2-3 DCV1-5 & 3	Exhausts blocked?	Y N ↓ 14	→ Test exhaust port fittings, silencers and flow controls	N Y ↓ 5	→ Remove blockage or replace part	Y ↓ 14
5 Start valve output?	Air at V1-2?	Y N ↓ 6	→ Test for blockages and check detent for correct spool position	N Y ↓ 6	→ Remove blockage, adjust detent or replace valve	Y ↓ 14
6 PV1 output? (not actuated)	Air at PV1-2?	Y N ↓ 7	→ Spring return effective? Valve stuck?	N Y ↓ 7	→ Unstick valve, replace spring or replace valve	Y ↓ 14
7 PV2 output?	Air at PV2-2?	Y N ↓ 8	→ Test for blockages or stuck poppet	N Y ↓ 8	→ Remove blockage or unstick/clean	Y ↓ 14
8 Air supply into DCV1?	Air at DCV1-1?	Y N ↓ 9	→ Remove supply pipework and test	N Y ↓ 9	→ Unblock supply	Y ↓ 14

Fig. 19.2 Fault, cause, remedy algorithm chart

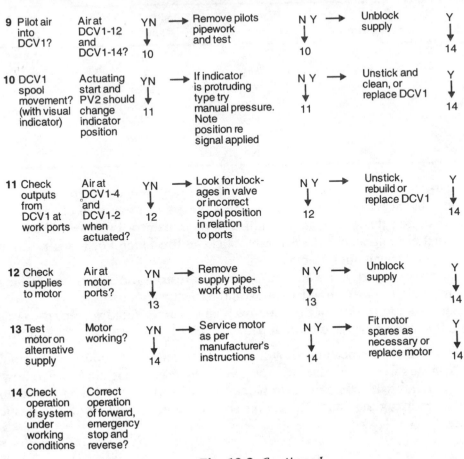

9 Pilot air into DCV1? Air at DCV1-12 and DCV1-14? Y N → 10 Remove pilots pipework and test N Y → 10 Unblock supply Y → 14

10 DCV1 spool movement? (with visual indicator) Actuating start and PV2 should change indicator position Y N → 11 If indicator is protruding type try manual pressure. Note position re signal applied N Y → 11 Unstick and clean, or replace DCV1 Y → 14

11 Check outputs from DCV1 at work ports Air at DCV1-4 and DCV1-2 when actuated? Y N → 12 Look for blockages in valve or incorrect spool position in relation to ports N Y → 12 Unstick, rebuild or replace DCV1 Y → 14

12 Check supplies to motor Air at motor ports? Y N → 13 Remove supply pipework and test N Y → 13 Unblock supply Y → 14

13 Test motor on alternative supply Motor working? Y N → 14 Service motor as per manufacturer's instructions N Y → 14 Fit motor spares as necessary or replace motor Y → 14

14 Check operation of system under working conditions Correct operation of forward, emergency stop and reverse?

Fig. 19.2 *Continued.*

Exercises for Chapter 19

1. The chart lists test points between each function through the circuit, and for a simple test 'yes, or 'no' for the presence of air at the point given, the chart helps to find the fault. What is the fault?
   2.9(c)

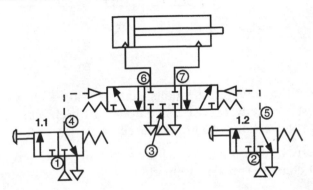

Test point	Valve 1.1 not actuated	Valve 1.1 actuated	Valve 1.2 not actuated	Valve 1.2 actuated
1	Yes	Yes	Yes	Yes
2	Yes	Yes	Yes	Yes
3	Yes	Yes	Yes	Yes
4	Yes	Yes	Yes	Yes
5	No	No	No	Yes
6	Yes	Yes	Yes	No
7	No	No	No	No

2. Called to replace a disintegrated piston seal on the system above in question 1, you find that seal failure has resulted in leakage across the piston of the actuator so that the cylinder will not extend. After the repair the cylinder will extend but a further fault occurs: the 5/3 valve sticks in the 'extend cylinder' position so the cylinder will not retract.
   (a) Indicate the part you would strip down to restore the system to working condition and explain what you would expect to find.
   (b) What additional item of preventative maintenance would you apply to this system after carrying out the repair?   2.9(c)

3. Due to a fault, in the position shown in the circuit air continuously leaks from port 5 of the directional valve. State (a) two likely reasons for the fault, (b) how, equipped only with some plugs to block the ports in the valve, you would go about identifying the fault. The answer should be in the form of a diagnostic routine.

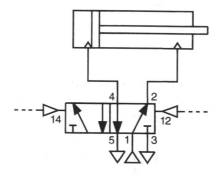

2.9(c)

4. For each of the following four faults, state two possible causes and for each cause suggest a remedy: (a) excessive noise, (b) high temperatures, (c) erratic operation, (d) hose, tube and pipe failures.   2.9(b)

5. Explain briefly how you would use the following as an aid to fault-finding: (a) a functional chart, (b) an algorithm chart.   2.9(c)

Appendix A
The syllabus

The following is the syllabus for City & Guilds 2340 Part 2 Certificate in Fluid Power Engineering (Pneumatics) – Component 002, Underpinning Knowledge.

Each element in the syllabus is identified by a coded objective and the corresponding pages in which it can be found in this book.

Objective

Pages

To demonstrate an understanding of pneumatic systems in general, candidates must prove an ability to:

2.1 State the purpose of pneumatic actuation as the means of providing linear/rotational power and list practical examples. ... 42–43

2.2 List the advantages/disadvantages of pneumatic systems compared to mechanical, electrical and hydraulic systems. ... 43–44

2.3 Describe the properties and behaviour of compressed air
  (a) state the concept of air pressure as force per unit area ... 9
  (b) distinguish between gauge pressure and absolute pressure ... 11
  (c) state, using a PV diagram, the relationship between pressure, volume, temperature and work done for isothermal, polytropic and adiabatic compression of air ... 13
  (d) define volumetric efficiency of a compressor ... 22
  (e) explain what is meant by 'free air delivered' (FAD) ... 16
  (f) explain what is meant by 'compression ratio' ... 23–24
  (g) define the term 'relative humidity' and explain the effect it has when air is compressed and when compressed air passes through a system ... 46–48
  (h) state the effects on air flow of changes in
    (i) temperature/density ... 50
    (ii) pressure difference ... 12–14, 49, 50, 52

2.6 Describe and prepare listed pneumatic circuits and associated methods of control

Appendix B:
A selection of practical tasks

As explained fully in Chapter 1, practical tasks are assessed according to the ability criteria, devised as a framework for the teaching and assessment of skill and split into the following four categories:

1.1 Interpret pneumatic circuit diagrams
1.2 Construct pneumatic systems from given information
1.3 Identify and rectify faults in pneumatic systems
1.4 Carry out routine maintenance on pneumatic systems.

The aim is to apply the required elements of competence and to prove that this has been done in the task report.

'1.1 Interpret circuits' and '1.2 Construct systems' may be combined in a task based on a circuit, such as those shown in this chapter or many of those in Chapters 10, 12, 13, 16 and 17.

'1.3 Fault-finding' may be carried out on a system constructed by another student for categories 1.1 and 1.2 which has then been intentionally damaged by inserting faulty components, introducing mechanical problems or altering pipework.

'1.4 Maintenance' may be a routine service of an item of plant, performance testing a line component, or stripping, servicing and testing an actuator, motor or compressor.

This is a short selection of circuits to help candidates develop their practical and diagnostic skills, raise confidence and improve their level of competence.

Fig. B.1 The 'cubic trainer'
(Photograph by kind permission of Technology Training Works.)

Note:
The procedure for carrying out this task is:
- with the flow regulators open move the
  cylinder to mid-stroke.
- close both flow regulator valves completely
  (air can enter the cylinder via the check valves).
In your documentation note each pressure gauge reading when the
FRL reads 4 bar: (1) with rod-end valve actuated, (2) with full-bore-end valve actuated.
Explain the reason for the readings observed.

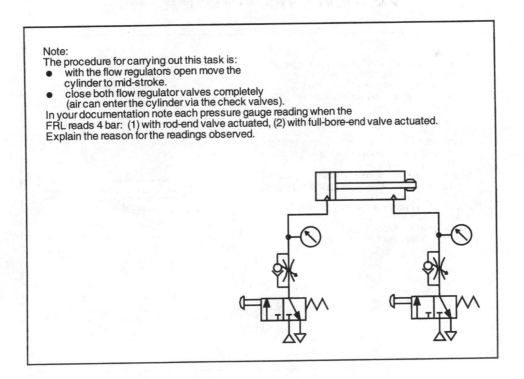

Task 1 Differential pressures

Note:
In your documentation of this task
explain the effect of putting equal
pressure to both ends of the cylinder.
(1) Give the reason for this effect.
(2) Suggest a change to the cylinder
      which would eliminate this effect.
Which part of the cylinder would you
measure to enable you to calculate
the force resulting from an input
pressure of 5 bar?

Any double-acting cylinder

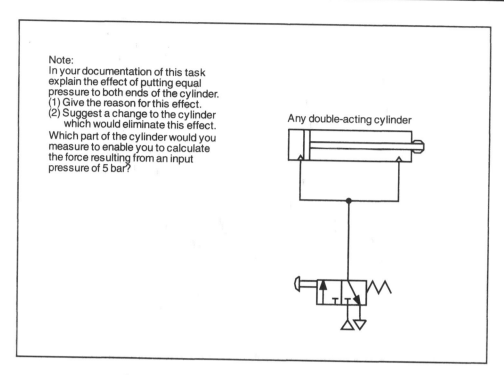

Task 2 Cylinder with both work ports pressurized

Note:
As drawn this circuit will give 'metering-in'
or 'metering-out' control only when the
cylinder is retracting. Start each test by
opening the valves and extending the
cylinder.

Carry out the following tests for supply pressures of
2 bar, 4 bar and 6 bar
(1) part close the rod-end flow control and fully
      open the full bore end flow control, to give
      very slow travel;
(2) part close the full-bore end flow control and
      fully open the rod-end flow control, to give
      very slow speed of travel.
Note how evenly or unevenly the speed
is controlled. Which mode of control,
'metering-in' or 'metering-out',
gives the better control of speed?

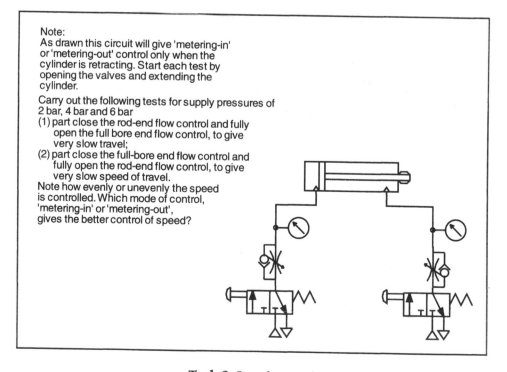

Task 3 Speed control

In your documentation note:
(1) the longest time delay obtainable
    by careful setting of the flow
    control on the timer circuit;
(2) the change in time delay if the
    main supply pressure is reduced
    to 3 bar.
Note how the 5/2 directional valve
is made to return by applying a
reduced opposing pressure on
the 12 pilot port of the valve.

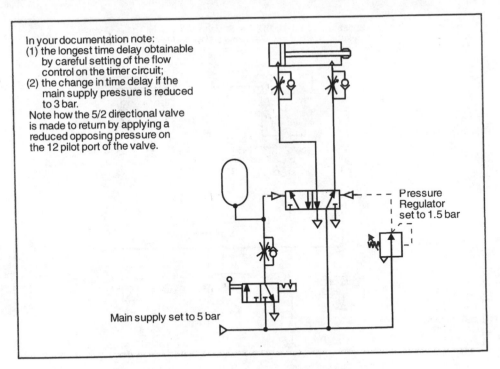

Task 4 Time delay and differential pilot

Complete and test the circuit with cylinder speeds
set to achieve a slow and even performance.
Draw a function diagram for the cylinders and control valves.

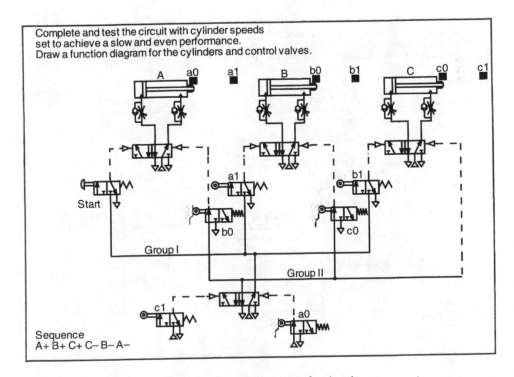

Sequence
A+ B+ C+ C− B− A−

Task 5 Two-group cascade circuit

A two-cylinder two-group cascade circuit with the start valve via an AND gate. Complete and test the circuit with cylinder speeds set to achieve a slow and even performance. Draw a function diagram for the cylinders and control valves.

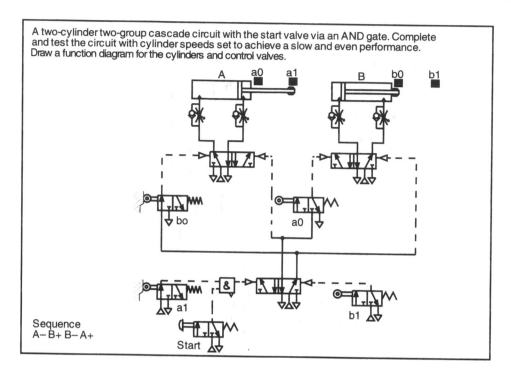

Sequence
A− B+ B− A+

Task 6 Two-group cascade circuit

A two-cylinder circuit with a switch valve and timer to give a delay when A is extended. Complete and test the circuit with the cylinder speeds set to achieve a slow and even performance. Draw a function diagram for the cylinders and control valves.

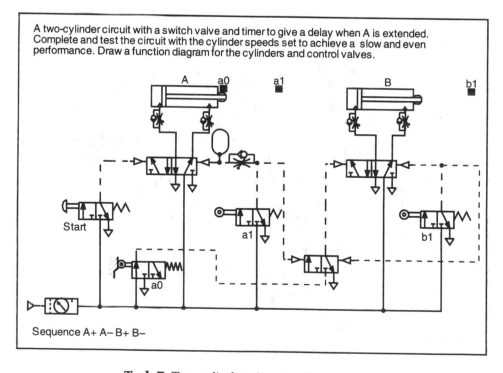

Sequence A+ A− B+ B−

Task 7 Two-cylinder circuit with time delay

A three-cylinder circuit which does not require cascade grouping. Complete and test the circuit with cylinder speeds set to achieve a slow and even performance. Draw a function diagram for the cylinders and control valves.

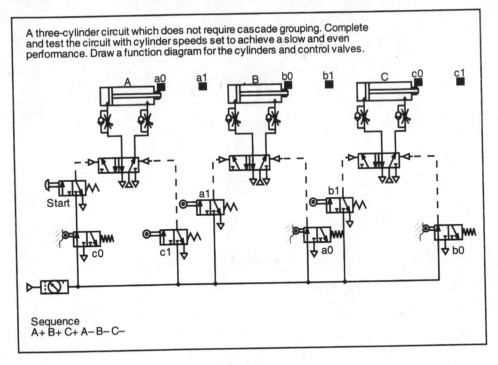

Sequence
A+ B+ C+ A− B− C−

Task 8 Three-cylinder circuit

Appendix C:
Symbols

Symbols 1

The symbols in this section generally conform to ISO1219-1.

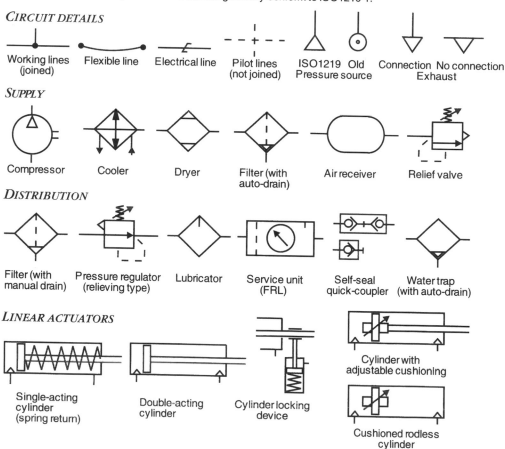

CIRCUIT DETAILS

Working lines (joined)   Flexible line   Electrical line   Pilot lines (not joined)   ISO1219 Pressure source   Old   Connection Exhaust   No connection

SUPPLY

Compressor   Cooler   Dryer   Filter (with auto-drain)   Air receiver   Relief valve

DISTRIBUTION

Filter (with manual drain)   Pressure regulator (relieving type)   Lubricator   Service unit (FRL)   Self-seal quick-coupler   Water trap (with auto-drain)

LINEAR ACTUATORS

Single-acting cylinder (spring return)   Double-acting cylinder   Cylinder locking device   Cylinder with adjustable cushioning   Cushioned rodless cylinder

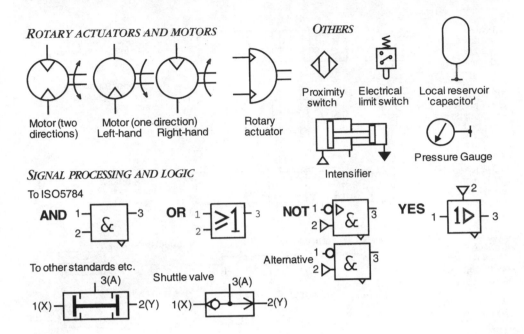

ROTARY ACTUATORS AND MOTORS

Motor (two directions)

Motor (one direction)
Left-hand   Right-hand

Rotary actuator

OTHERS

Proximity switch

Electrical limit switch

Local reservoir 'capacitor'

Intensifier

Pressure Gauge

SIGNAL PROCESSING AND LOGIC

To ISO5784

AND   1 — & — 3   2

OR   1 — ≥1 — 3   2

NOT   1 — & — 3   2

YES   1 — 1▷ — 3   ▽2

Alternative   1 — & — 3   2

To other standards etc.

3(A)
1(X) — — 2(Y)

Shuttle valve
3(A)
1(X) — — 2(Y)

Symbols 2

The symbols in this section generally conform to ISO1219-1

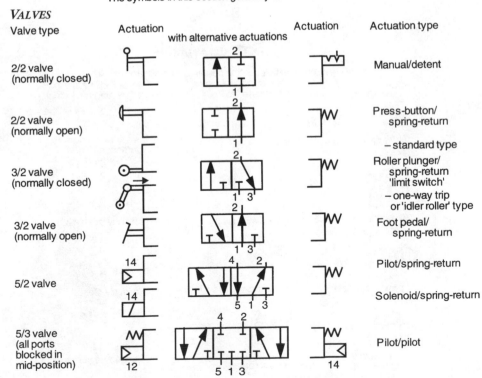

VALVES

Valve type	Actuation	with alternative actuations	Actuation	Actuation type
2/2 valve (normally closed)		2 / 1		Manual/detent
2/2 valve (normally open)		2 / 1		Press-button/ spring-return
				– standard type
3/2 valve (normally closed)		2 / 1 3		Roller plunger/ spring-return 'limit switch'
				– one-way trip or 'idler roller' type
3/2 valve (normally open)		2 / 1 3		Foot pedal/ spring-return
5/2 valve	14 / 14	4 2 / 5 1 3		Pilot/spring-return
				Solenoid/spring-return
5/3 valve (all ports blocked in mid-position)	12	4 2 / 5 1 3	14	Pilot/pilot

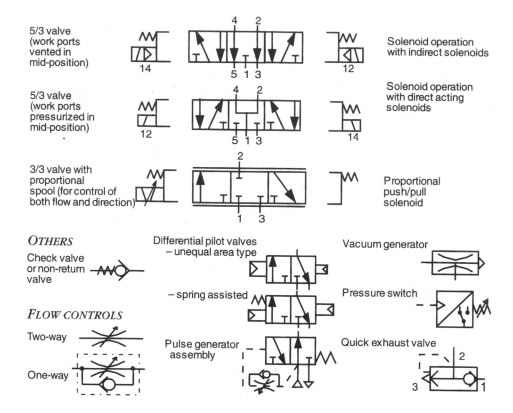

5/3 valve (work ports vented in mid-position)

Solenoid operation with indirect solenoids

5/3 valve (work ports pressurized in mid-position)

Solenoid operation with direct acting solenoids

3/3 valve with proportional spool (for control of both flow and direction)

Proportional push/pull solenoid

OTHERS

Check valve or non-return valve

Differential pilot valves – unequal area type

Vacuum generator

– spring assisted

Pressure switch

FLOW CONTROLS

Two-way

One-way

Pulse generator assembly

Quick exhaust valve

Electrical Symbols

The symbols in this section generally conform to DIN40713.
Note: American electrical symbols, a common option for PLC program diagrams, are shown and referred to in Chapter 17 as 'ladder diagram format'.

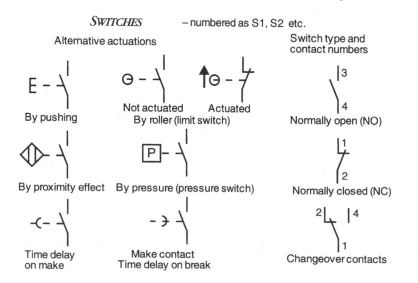

SWITCHES – numbered as S1, S2 etc.

Alternative actuations

Switch type and contact numbers

By pushing

Not actuated   Actuated
By roller (limit switch)

3

4

Normally open (NO)

By proximity effect   By pressure (pressure switch)

1

2

Normally closed (NC)

Time delay on make

Make contact
Time delay on break

2   4

1

Changeover contacts

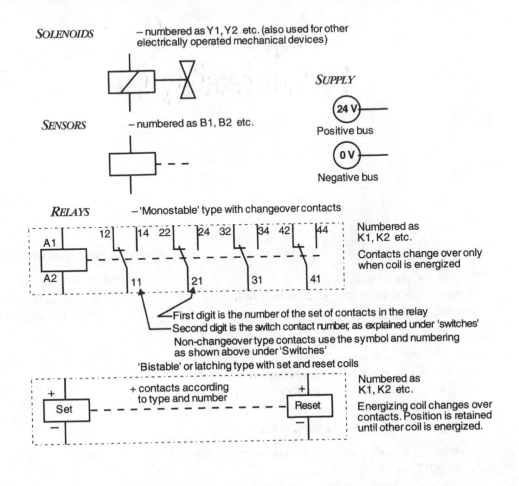

SOLENOIDS — numbered as Y1, Y2 etc. (also used for other electrically operated mechanical devices)

SUPPLY

24 V
Positive bus

SENSORS — numbered as B1, B2 etc.

0 V
Negative bus

RELAYS — 'Monostable' type with changeover contacts

A1 | 12 | 14 | 22 | 24 | 32 | 34 | 42 | 44 | Numbered as K1, K2 etc.
A2 | 11 | 21 | 31 | 41 |

Contacts change over only when coil is energized

First digit is the number of the set of contacts in the relay
Second digit is the switch contact number, as explained under 'switches'
Non-changeover type contacts use the symbol and numbering as shown above under 'Switches'

'Bistable' or latching type with set and reset coils

+ contacts according to type and number +
Set   Reset
−   −

Numbered as K1, K2 etc.

Energizing coil changes over contacts. Position is retained until other coil is energized.

Appendix D
Further reading list

Title and description	Cost	Publisher or source
The principles of Circuit Design A compact summary of typical simple industrial circuits.	£2.00	Compair Maxam Ltd, Carn Brea, Redruth, Cornwall TR15 3PR
Engineering Applications of Pneumatics & Hydraulics by Ian C Turner. ISBN 0 340 62526 0 Assuming only basic knowledge and with a strongly practical bias, this book aims to equip the reader with a sound understanding of fluid power systems and their uses in practical engineering.	£12.99	Edward Arnold (Publishers) Ltd.
Pneumatic Handbook – 8th Edition edited by Barker. ISBN 1 85617 249 X A standard reference work on modern pneumatic and compressed air engineering, this book provides an essential up-to-date guide for users of pneumatic systems.	£135.00	Elsevier Advanced Technology
Pneumatics, Basic Level Textbook by Peter Croser. ISBN 3 8127 3131 2. Order No. 93131 This course teaches the basic physical properties of pneumatics and the function and application of Pneumatic components.	£33.80	Festo Ltd, Automation House, Harvest Crescent, Ancells Business Park, Fleet, Hants GU13 8XP
– Basic Level Excrcises by D Walter. Order No. 94001	£22.35	As above
Fundamentals of Pneumatic Control Technology by Hasebrink, Kobler & Idler. ISBN 3 8127 0851 5. Order No. 90851. The book provides an introduction to the general fundamentals of control technology, explains	£29.98	As above

symbols, defines terms and basic control systems
and goes on to deal with the systematic drafting of
control systems.

Electro-Pneumatics Basic Level by Croser & Thomson.
ISBN 3 8127 1181 8. Order No. 91181.   £36.13   As above
   General control technology, motion and switching,
fundamentals of electrics/electronics, components,
safety, electrical symbols and circuit diagrams.
   - Basic Level Exercises by G Fiedler. Order No.   £33.19   As above
93036.

Programmable Logic Controllers, Basic Level   £27.02   As above
Textbook by R Ackermann. ISBN 3 8127 3311 0.
Order No. 93311.
   Control systems used in conjunction with sensors,
processors and actuators, also programming
methods. With exercises and a collection of
practical examples.
   - Basic Level Exercises by R Ackermann. Order No.   £81.07   As above
93306.

Maintenance of Pneumatic Equipment & Systems   £35.39   As above
Textbook by Meixner and Behr. ISBN 3 8127 0841 8.
Order No. 90841.
   Intended for servicing and installation personnel,
including air treatment, current symbols, basic
circuits, functional sequences and the assembly of
control systems, with practical examples and
guidance on fault-finding.

Festo Ltd. also list Advanced Level textbooks,
exercises and supporting material together with
Training Equipment.

A Course in Applied Pneumatics (Training Manual).   £29.17   IMI Norgen Ltd.,
   This updated Manual deals with most aspects of   Training Dept.,
the City and Guilds 2340 Part 2 syllabus up to the   Brookside Business
level of cascade circuits.   Park, Greengate,
Middleton,
Manchester M24 1GS

The Pneumatic Trainer Vol. 1.   £35.00   Mannessmann Rexroth
   The basic principles of pneumatic components,   Ltd., Technical Services
with excellent illustrations and clear explanations.   Dept., Cromwell
Road, St. Neots,
Cambs. PE19 2ES

The Pneumatic Trainer Vol. 2. Electro-pneumatics. This new book extends the basic training into the area of combining electrics with pneumatics.	£35.00	As above
Pneumatic Control for Industrial Automation – Textbook by Peter Rohner. ISBN 0 471 33463 4. Aimed to support courses in fluid power, industrial automation and pneumatic control, this book explains the function, operation of and typical applications for valves, timers, actuators, air motors, compressors, air preparation and air distribution components.	£34.95	John Wiley Publishers, Chichester
Manual of Pneumatic Systems Optimization by Henry Fleischer. ISBN 0 07 021240 6. Ways to minimize component size, boost productivity, optimise compressed air usage and reduce utility bills etc., including: The Evolution of Pneumatic-System Technology, Conductance, Actuators, Fittings, Conductors, Valves, Accumulators, Receivers, Reservoirs, Surge Tanks and Vessels, Air Motors, Flow Controls, and Quick-Exhaust Valves. Other Pneumatic Components and Parameters, Pneumatic Venting.	US$49.50	McGraw-Hill Inc.
Basic Pneumatic Control Technology (Parker Part No. TRGGB-3). Covers the Evolution of pneumatics, Health & Safety, Theory, Terminology, Air compression, Preparation and Distribution, Cylinders, Directional controls, ancillary equipment and basic circuits.	£24.00	Parker Hannifin plc, or (entitled 'Airways') from Economatics Educational, Epic House, Darnall Road, Attercliffe, Sheffield S9 5AA
Power Pneumatics by Brian Callear and Michael J. Pinches. ISBN 0 13 489790 0. A comprehensive guide to pneumatic components and systems, with explanations and solutions for many of the field's most important problems. An up-to-date guide to power pneumatics system design, component selection, and problem solving.	£21.95	Prentice Hall Professional Technical Reference
Pneumatic Training Course PN/S1 and PN/K1. A structured introduction to components and circuits up to the level of stepper circuits, with exercises.	£10.00	Robert Bosch (Pneumatics) Ltd., Meridian South, Meridian Business

		Park, Braunstone, Leicester LE3 2WY
Pneumatic Theory and Applications. A basic introduction to Pneumatics up to the level of sequence control, also basic electro-pneumatic control.	£25.00	As above
Pneumatic Position & Pressure Servo Proportional Control. A structured series of exercises in ring binder format.	£25.00	As above
Pneumatic Control for Industrial Automation – Workbook 1 by Peter Rohner. ISBN 0 646 00565 0.	A$24.95	Rohner Fluid Power Automation & Control Pty. Ltd., 14 Turner Street, Briar Hill, Vic. 3088, Australia
Also available in the same series: Teacher Book 1 Constructional exercises.	A$28.95 A$19.95	
Pneumatic Technology (Training Manual and Workbook). Dealing with the applications, theory, compression, treatment and distribution of air, together with components and circuitry up to cascade systems, this book is well illustrated and clearly explained. The Workbook includes symbols and 25 practical exercises.	£42.36	SMC Pneumatics Ltd., Vincent Avenue, Crownhill, Milton Keynes, MK8 0AN, or SMC Fast Response, Freepost, Aylesbury, Bucks HP22 5BR
Electro-pneumatic Technology (Training Manual and Workbook). Electrical principles and electro-pneumatic components are explained, together with circuits employing relays. The Workbook includes symbols and 16 practical exercises.	£26.02	As above
The Pneu Book. An informative reference manual with sections covering issues from the sizing of pneumatic valves and cylinders, guides to terminology, circuit design, symbols, vacuum theory and cylinder efficiency.	FOC	As above
Compressed Air Course (8 Folios). Uses and properties of air, Compressors, Air treatment and distribution, Humidity, Actuators and motors, Pipe sizing, Force calculations, Valves, Simple circuits, etc.	FOC	Spirax Sarco Ltd., Charlton House, Cheltenham, Glos. GL54 5EL

Computer based learning material

Pneusim Pro. Circuit design and simulation software. An effective PC-Windows CAD program which can also simulate the working of a system on a continuous or step-by-step basis. Numerous prepared circuits are included plus symbol libraries enabling almost any pneumatic system to be drawn and tested on-screen. Includes ladder logic diagrams in European and US format and also digital electronics.   £211.00   IMI Norgren Ltd., UK Training Dept., Brookside Business Park, Greengate, Middleton, Manchester M24 1GS

CD-ROM with the following sections, supported by an explanatory booklet.
Compressed Air Production, Preparation, Distribution and Safety, Introduction to Pneumatic equipment, Circuit design/Basic circuits, Electro-pneumatics, Logic, Maintenance and fault-finding, Applications, Database/Glossary, Test module.   £350.00   Parker Hannifin plc, Pneumatic Division, Walkmill Lane, Bridgtown, Cannock, Staffs WS11 3LR

Index